ENERGY POSSIBILITIES

SUNY Series in Science, Technology, and Society
Edited by Sal Restivo and Jennifer Croissant

ENERGY POSSIBILITIES

*Rethinking Alternatives
and the Choice-Making Process*

Jesse S. Tatum

STATE UNIVERSITY OF NEW YORK PRESS

The following publishers have generously given permission to use quotations from copyrighted works. Reprinted from *Energy*, vol. 17, Jesse S. Tatum, "The Home Power Movement and the Assumptions of Energy Policy Analysis," pp. 99–108, © 1992, with kind permission from Elsevier Science Ltd., The Boulevard, Langford Lane, Kidlington OX5 1GB, UK. From *The Whale and the Reactor* by Langdon Winner. © 1986 by the University of Chicago. Reprinted by permission of The University of Chicago Press. From "Energy Strategy: The Road Not Taken," *Foreign Affairs*, by Amory Lovins. © 1976 by the Council on Foreign Relations, Inc. Reprinted by permission of *Foreign Affairs*, (October 1976). From *Energy in Transition: 1985–2010*, National Academy of Sciences, 1979. © 1979 by the National Academy of Sciences. Reprinted with permission. From *Sociopolitical Effects of Energy Use and Policy*, National Academy of Sciences, 1979. © 1979 by the National Academy of Sciences. Reprinted with permission. From "Energy in the U.S. Economy," *Science 225* (August 1984), p. 890–897 by Cleveland, Costanza, Hall, and Kaufmann. © 1984 by the AAAS. Reprinted by permission. From *Technology and the Character of Contemporary Life* by Albert Borgmann. © 1984 by the University of Chicago. Reprinted by the permission of The University of Chicago Press. Reprinted with the permission of The Free Press, a division of Simon & Schuster from *The Informed Heart: Autonomy in a Mass Age* by Bruno Bettelheim. Copyright © 1960 by The Free Press; copyright renewed 1988 by Bruno Bettelheim. From "Technology and Values: Getting Beyond the Device Paradigm Impasse," Jesse S. Tatum, *Science, Technology, & Human Values*, (19:1) pp. 70–87, copyright © 1994 by Sage Publications, Inc. Reprinted by permission of Sage Publications, Inc.

Published by
State University of New York Press, Albany

© 1995 State University of New York

For information, address State University of New York Press,
State University Plaza, Albany, N.Y., 12246

Production by Cathleen Collins
Marketing by Fran Keneston

Library of Congress Cataloging in Publication Data

Tatum, Jesse S., 1952–
    Energy possibilities : rethinking alternatives and the choice-
making process / Jesse S. Tatum.
       p.   cm. — (SUNY series in science, technology, and society)
    Includes bibliographical references and index.
    ISBN 0–7914–2595–9. — ISBN 0–7914–2596–7 (pbk.)
    1. Power resources.   2. Energy policy.   I. Title.   II. Series.
TJ163.2.T347   1995
333.79—dc20   94–37133
                                                 CIP

10 9 8 7 6 5 4 3 2 1

Contents

Preface

In writing this book, I am compelled by a belief that present energy policies and practices in the Western world do not reflect the best interests of ordinary people and by a sense that the decision making processes by which they have been selected fall far short of our best democratic traditions and aspirations. My purpose here is to offer the reader a new perspective on the energy problem, to argue that traditional analytical methods for arriving at solutions may well be both inadequate and inappropriate, and to suggest that the real question posed by the energy problem is one of how the decision maker (i.e., the reader) wishes to live in the world. The reader should be forewarned that, after redefining the problem, I will not be offering "The Solution." The book does move from theoretical considerations into a discussion of unique practical responses to energy concerns, and an effort will be made to fortify the reader's thinking by drawing what guidance is possible from both theory and practice. But the intent here is more to place the problem in the reader's lap, where it belongs, than it is to offer any closed solution.

The book begins (Chapter 1) by raising broad questions about the validity and wisdom of traditional energy policies and policy methods and of individual and collective responses to the energy challenge. The significance of these questions is reinforced in Chapter 2 through consideration of the profound influence of energy choices in shaping who we are and how we live. It is then argued more specifically (Chapter 3) that present policies and methods are neither as "disinterested" nor as "objective" as is generally supposed, given a widespread failure to identify and examine underlying assumptions and commitments. Careful analysis of traditional engineering and economic perspectives is developed or reviewed to reveal apparent logical fallacies and highly debatable assumptions that are not popularly recognized. Chapter 4 reviews a collection of theoretical arguments that compete with the accepted view, including interpretations that focus on the physical rather than the economic role of energy in material production. Although these competing views appear to be held only by a minority in the policy community, it must be noted that "reality" is not determined by an opinion poll of designated experts. The reader is left with the author's concern that alternative interpretations may be of greater practical and theoretical value than is ordinarily recognized, a possibility that raises grave doubts about existing energy policies.

After dealing with energy issues from these theoretical perspectives, an interpretation of the practical origins of present energy policies and practices is offered (Chapter 5), suggesting that these are neither the outcome of truly democratic processes nor the product of definitive technical calculations. Reflecting established distributions of interest more than anything else, our behavior appears to be the product of a two-pronged default: present policies and practices arise primarily from a combination of institutional or collective momentum, on the one hand, and, on the other, from individual or popular nonparticipation. This position is defended through brief analysis of the energy policy debates of the 1970s and 1980s, including the surprise entry of Amory Lovins's notion of "soft energy paths," and through other practical and theoretical arguments.

Turning finally to a specific example of alternative practice, I describe (Chapter 6) the recent history of the "home power" movement in which growing numbers of participants have adopted their own renewable electricity and other energy systems in spite of higher costs than those characteristic of conventional energy sources. The home power experience offers concrete confirmation of the clearest conclusion from preceding chapters: the range of possibility, both technically and in terms of social, political, or cultural configuration, is far more extensive with respect to energy choices than the stunted collection of "energy futures" given serious consideration by policy makers even twenty years after the "energy crisis." If choices are to be made democratically within this broadened range of possibility, it is argued, a more experimental (even "playful") approach, exemplified by home power efforts, will be essential in first collectively exploring that range of possibility. A process more open to the subtleties of experience and to diverse and unique perspectives may also be in order as energy systems are chosen not because they are the cheapest source of British thermal units (BTUs) or kilowatt-hours (kWhs), but because they are valued as defining elements of satisfying relationships with other people and with the natural environment.

The central purpose of this book is to offer a different, if not entirely new, perspective on energy issues and the means by which they are, and are to be, resolved. This endeavor seems more than justified in view of the pervasive importance of energy decisions in shaping the way we live. I would like to believe that the perspective offered here is the product of long, careful, and more than usually penetrating thought. I further console myself with every conscious effort to communicate honestly in the belief that my own values are best served by honestly offering a perspective for the reader's consideration, rather than by trying to win in a debate through hidden maneuvering or selective presentation with deceptive intent. At some recess of my mind, however, I must undoubtedly also have an interest in persuasion. At risk of alienating the reader at the very start, therefore, let me note at the outset that I am firmly convinced that a variety of energy alternatives, especially alternatives that combine renewable (solar) energy sources with what conservation advocates themselves would generally dismiss as "extreme" efficiency measures, deserve much more careful consideration than they have received to date. The reader is left to assess any expositional "bias" that may be introduced by this conviction as it may be reflected in the structure and detail of the book. Please do not dismiss as bias, however, what is simply a *departure* from conventional wisdom, while endorsing as objective

analysis what is merely *consistent* with the status quo. To closely paraphrase Herman Daly, the fact of having arrived at a definite conclusion is not by itself evidence of bias, any more than the appearance of
not having arrived at a conclusion is evidence of lack of bias—the latter could, among a number of other possibilities, reflect a simple lack
of conscientious thought.[1]

It should be noted that the perambulations of this book are ultimately speculative in nature. The issues raised go beyond proof as, I
suspect, does any matter of real human significance. While any number of facts may inform our decision making, energy matters are far
too complex to be resolvable from facts alone. Simply bounding the
discussion, whether to economic comparisons, for example, or more
inclusively to consider the social imperatives of particular technology
choices, implies an exercise of human judgment or practical politics.
Proof becomes impossible. *Action*, on the other hand, arises quite
properly from beliefs, values, and careful judgment, whatever their
particular relation to facts. This is true with respect to energy issues, I
would assert, no matter what position, traditional or radically innovative, one chooses to adopt. My objective here is not proof but, again,
communication of an unconventional perspective on energy issues
and their handling as matters of public and individual policy.

What has been said so far should already foreshadow one other
message that I hope will be evident throughout this book, bearing on
the distinction between a concern with *outcomes* and a concern with
process. Energy, environmental, and other policy discussions are often
focused on material outcomes. In the energy arena, we tend to focus
on the range of resources and conversion technologies that might be
employed and on a variety of production and consumption scenarios
for the future. Particular notions of "realistic" expectations are then
used as levers to effect "appropriate" changes. Without questioning
the obvious significance of practical outcomes, this Preface and the
text that follows are intended to be expressive of a somewhat different
focus of concern. In a sense the real issue, after all, is one of process.
We must also be concerned, that is, with the underlying human relationships that are expressed or instituted through the process of making choices that are of fundamental human significance. I do believe

that energy and related social and environmental decisions will be of pivotal importance in determining our future survival and the sustainability of chosen patterns of material production; but outcomes, even in this league, seem to me to pale in importance when compared with concerns about the fairness and legitimacy of the decision making processes and the degree to which those processes elevate or degrade each of us while we may live. Material outcomes, after all, are ultimately beyond our complete control. It is only in how we treat each other, individually and through the arrangement of social and political relations, that we may hope truly to make, and bear full responsibility for, our own choices. Outcome and process, both, are of legitimate concern; the latter is given relatively greater weight here, I hope, than is the custom.

To place this work in its proper academic setting, it may be worth noting that the perspective offered owes much to the emerging discipline of "science, technology, and society" (STS) studies. The book as a whole could even be thought of as an exercise in "constructive technology assessment."[2] In keeping with the motivating concerns of the book and my desire to reach more than an academic audience, however, I have restricted reference to the academic literature largely to the footnotes. A familiarity with STS as a field is not assumed.

One final note on style. The majority of those who read my work tell me that I overuse "waffle words" like seems, appears, suggests, and I think or I believe. Such terms come naturally to me in dealing with significant realities that remain, for me, unknown and ultimately unknowable. Bending to the will of the majority, however, I have made some effort to restrict my use of such terms. I do, nevertheless, regard virtually everything I have written here as tentative. My only claim to "expertise" is having spent a great deal of time over the last twenty years reading, thinking, discussing, and writing about energy issues and about other points at which science, technology, and society intersect. It will remain for the reader to decide the relevance of these pursuits and whether they add or detract from the weight of what is said here. Disciplinary professionals (economists, engineers, etc.) may not share my sense of uncertainty, supported and surrounded, as they are, by legions of like-minded professionals in their own disciplines. For others, it may simply go without saying that what an author puts in a book is simply what that author "thinks." Still, I would prefer to qualify each of the grand pronouncements of this book: "this is the way the matter appears to me now, as best I can tell."

Acknowledgments

In reflecting on the years of effort that have led to this book, I must express my warm regard and appreciation to the faculty and especially to the graduate students of the Energy and Resources Group at the University of California at Berkeley. Beyond this I will not try to list the countless excellent teachers and other dedicated helpers whose efforts are in some way reflected here. Only a small fraction of these are specifically recognized in referring to particular published works. I owe special thanks to John Holdren, Ted Bradshaw, Derrick Tucker, Joe Eto, Mark Christensen, and Bart McGuire. I am most grateful, however, to my mother and father and to Langdon Winner, Herman Daly, David Noble, Laura Nader, and Myles Horton; without the inspiration of their unique and uncompromising integrity, this project would not have reached the stage of conception.

ONE

Raising the Question

On a number of occasions in the late 1970s while the author was working at Oak Ridge National Laboratory, he found himself engaged in the almost ritualized exchange of air travelers in which passengers in adjacent seats explore each other's employment, destinations, and cities of origin. On hearing that I was doing work on renewable energy systems at Oak Ridge, and apparently taking me for one of those elusive "experts" to whom we so often hear reference, an unexpected spark of more than ritualistic interest would often appear and my fellow traveler would ask; "Are solar energy systems really cost competitive?" or, perhaps, "When are those solar energy systems going to become cost competitive?"

These were very awkward questions for someone fresh out of graduate school and much steeped in concerns about "externalities," especially the environmental effects of energy production that are not "priced" or otherwise taken into account in ordinary market exchanges. (The air pollution from coal combustion for electric power production would be an example of an "externality" because the electricity user traditionally does not pay for the costs of that pollution in his or her monthly electric bill.) In making energy and other choices, "cost competitiveness" can be a complex and ambiguous standard. My fellow passengers were asking what they clearly took to be simple, direct, and even technical questions hoping, perhaps, for a more simple, direct, and unbiased answer (from a real live expert) than they felt they had been able to cull from conflicting media and political exchanges. The standard answer to these questions at the time (perhaps still today) was that except for certain residential space and water heating applications, solar energy systems were not, and were not soon likely to be, cost competitive with other sources of energy. But giving the standard answers left the potentially critical issue of the appropriateness of the questions themselves unaddressed.

Awkward from the start, these little airline exchanges became increasingly disturbing. How was it that everyone seemed to be asking the same one or two questions? Whose questions were these anyway? Were they in fact arising spontaneously and independently as the only logical issues to be probed?

Even more disturbingly, whose responsibility was it to decide whether or not these were the appropriate questions to ask? Was this my job as a public servant, or was this the responsibility of those asking the questions? Were the answers to these apparently simple questions going to be determinative in shaping the energy choices of the individuals I had talked with—or determinative in the selection of an energy future for society as a whole? And if so, who was responsible for those decisions?

How were we to know whether or not a "cost competitive" energy system was the best or wisest choice for the individual or for society? Who was to determine the degree to which "cost competitiveness" and "wisdom" overlapped?

Most people, airline passengers or otherwise, probably do not feel they have time for the complications and confusion introduced by this enumeration of questions about the question (nor was I, at the time, at all equipped to define, let alone suggest answers to, most of these concerns). This fact in itself, however, is suggestive of the influence of processes of socialization and social control—and of the possibility of related failures of omission or commission in the exercise of responsibility for choosing energy futures—that were, perhaps, the origin of my discomfort.

Social Control?

In his book, *The Informed Heart: Autonomy in a Mass Age*, psychologist Bruno Bettelheim[1] draws on his experience as a prisoner in a Nazi concentration camp to provide starkly focused illustrations of a variety of mechanisms of social control and to comment in often less explicit terms on modern life in a "mass age." Reading this book some years after leaving Oak Ridge National Laboratory to return to graduate study, I was often reminded of the exchanges I had had on airplanes regarding the cost competitiveness of solar energy systems. In reading the following passages, for example, I was reminded of the feeling I had had that I was actually being asked by my airplane interlocutors what they should do (a question I felt I had no business answering).

> When social change is rapid, there is not enough time to develop the new attitudes needed for dealing with an ever changing environment in terms of one's own personality. This makes the individual "confused" and uncertain. The more this happens, the more he watches to see how others meet the new challenge and tries to copy their behavior. But this copied behavior, not being in line with his own make-up, weakens his integration and he grows less and less able to respond with autonomy to new change.
>
> What we now fear is a mass society in which people no longer react spontaneously and autonomously to the vagaries of life, but are ready to accept uncritically the solutions that others offer; we fear also that those solutions are geared only to technological progress, disregarding the greater integration it requires.
>
> . . . It is hard to say where exactly in this evolution of the mass state we now stand.[2]

To what degree were those asking the cost competitiveness question actually seeking new information and to what extent were they asking, as a child might more appropriately ask, "What should I do in this situation?" Did the questions they asked stem from their own "integrated" personalities or were these the questions their neighbors

asked? Were they indicative of "autonomous" reactions to new technical possibilities or did they arise from an acquiescence to a mass age
and an abandonment of efforts to deal with "the vagaries of life" from
an integrated base of personal principles, values, and attitudes?

At another point in his book, Bettelheim describes one of many
extreme forms of social control employed in the concentration camps.

> Among the worst mistakes a prisoner could make was to watch
> (to notice) another prisoner's mistreatment. There the SS seemed
> totally irrational, but only seemed so. For example, if an SS man
> was killing off a prisoner and other prisoners dared to look at
> what was going on in front of their eyes he would instantly go
> after them, too. But only seconds later the same SS would call the
> same prisoners' attention to what lay in store for anyone who
> dared to disobey, drawing their attention to the killing as a warn-
> ing example. This was no contradiction, it was simply an impres-
> sive lesson that said: you may notice only what we wish you to
> notice, but you invite death if you notice things on your own voli-
> tion. . . . the prisoner was not to have a will of his own.
>
> . . . To know only what those in authority allow one to know is,
> more or less, all the infant can do. To be able to make one's own
> observations and to draw pertinent conclusions from them is
> where independent existence begins. To forbid oneself to make
> observations, and take only the observations of others in their
> stead, is relegating to nonuse one's own powers of reasoning, and
> the even more basic power of perception.
>
> . . . Knowing that . . . an emotional reaction [to mistreatment of
> prisoners] was tantamount to suicide, and being unable at times
> not to react emotionally when observing what went on, left only
> one way out: not to observe, so as not to react. So both powers,
> those of observation and of reaction, had to be blocked out vol-
> untarily as an act of preservation. But if one gives up observing,
> reacting, and taking action, one gives up living one's own life.
> And this is exactly what the SS wanted to happen.[3]

Analogies between such extreme forms of social control and possible
social control of attitudes regarding energy alternatives may seem too

farfetched to consider. Bettelheim's description makes the particular mechanism employed here very clear, however. In its blandest form in the context of energy issues, one might ask, "Are all spontaneous reactions to energy alternatives received equally in society?" Is an inchoate or even a cogently articulated noneconomic attraction for renewable energy alternatives taken to be valid in our society, or are certain "rational" reactions more acceptable than other "less rational" reactions? Were some of those asking the question on the airplane (e.g., those who seemed to take a "more than ritualistic" conversational interest in my work on renewables) actually seeking *permission* for their spontaneous sense of attraction for renewable alternatives? Or, having already accepted others' definitions of what observations are and are not pertinent and what logic is to be applied in proceeding from observation to choice of action, were they simply hoping that someone would go through the final motions also—that is, assess the costs and energy outputs and apply the requisite (economic) logic—for them? Although the sanctions are surely less extreme in this case, is the question asked on the airplane not indicative of a similar abandonment of independent capacities for observation and reaction?

Precedence of Economic Argument

While there is room for debate over the degree to which economic calculations actually shape individual and social energy decisions, the established precedence of economic over other forms of argument seems beyond question. One of the clearest statements of this precedence has been provided by a National Research Council study, *Energy Use: The Human Dimension*,[4] completed in 1984. Its authors begin by describing four logical alternative views of energy. There is the "commodity" or economic view, under which energy consists simply of "energy forms or energy sources that can be developed and sold to consumers." But energy can also be viewed as an ecological resource. Under this view, energy sources are classified as renewable or nonrenewable and the effects of extraction and use on soil, water and air quality, on climate, and on biological communities are taken into consideration. As another possibility, energy can be viewed as a social necessity; according to this view, energy for cooking, lighting, home heating, transportation, and other essential uses is taken first to be a basic right in any just and equitable society. Finally, the study authors describe a view of energy as a strategic material. If one adopts this view, the critically important attributes of energy sources become geographical location, political stability and orientation of source nations, and the availability of domestic or other reliable substitutes in the event of a cutoff of specific sources. While each of these views has logical appeal and rests on a set of values with broad if not universal appeal, the commodity view is dominant.

> The four basic conceptions of energy do not have equally strong support, either in the political arena or among policy analysts. In most aspects of the national policy process, the commodity view is dominant. Dominance of a particular view of energy does not mean that it is the only view given consideration, but that other views must make special claims before being taken seriously. And in most U.S. energy policy debates, the burden of proof still remains on those who assert that energy should be treated as something other than an ordinary commodity. When these advocates succeed, they do so by winning exceptional treatment for particular situations rather than by changing the dominant perspective.[5]

All reactions to our energy alternatives clearly are *not* treated equally. Surely one would have to expect summary dismissal, in fact, of any reactions not yet in a form as cogently communicated as the preceding ones when even the latter face an uphill battle in any struggle for exceptions to the dominance of the commodity view. Rightly or wrongly, instincts, intuitions, aesthetic preferences—if indeed they were ever seriously advanced in policy making settings—would be dismissed as impractical indulgences, not even to be explored in the face of the cold hard "facts" of economic analysis.[6]

That our attitudes and reactions to particular energy alternatives are shaped by processes of socialization and by a variety of forms of social control seems beyond denial and certainly should come as no great surprise, given the ubiquitousness of the processes involved. These processes seem so "natural," for the most part, in fact, that they are nearly transparent,[7] calling attention to themselves, if at all, only in exceptional cases when they become explicit or begin to be overtly coercive. In his preface to the NRC study referred to earlier, for instance, the study chairman applauds the Department of Energy (which funded the study) for recognizing the potential of the noneconomic social sciences for informing energy policy. But he also quotes from DOE's charge to the committee defining the committee's purpose and task: "*Economic paradigms*, together with assessments of the potential contributions of new and existing technologies, *will continue to provide the basis for the analysis of alternative public policies* relating both to energy production and consumption. At the same time, there is considerable evidence to suggest that the noneconomic behavioral and social sciences can contribute significantly to such analyses."[8] This charge clearly states that economic (and engineering) views are and will remain dominant and that the behavioral and social scientists who will be conducting the study being commissioned must restrict their investigations to ways in which they might contribute within the framework established by the dominance of these views. While this position may be generally implicit in research of this sort, such an explicit restriction in the commissioning of academic research may give us pause. Socialization is one thing; more or less explicit instructions as to permissible and impermissible avenues of thought and investigation may be quite another.

Effective Conspiracy and Other Questions

I do not mean to suggest any sort of clearly defined conspiracy here. Nor do I intend to suggest that economic treatments of energy options are necessarily unwise or inferior to other views of these issues, whether the latter are cogently stated and widely supported or as yet inchoate in the minds of a handful of individuals. One must surely be "practical" at some level about energy matters in a sense that attends to the "affordability" or "sustainability" aspects of "economic" considerations, if not to classical cost competitiveness.

Any discussion of processes of socialization or social control does, however, raise another important collection of questions about who stands to benefit and who stands to lose from the results of the process. It may be that any society is identifiable as such only because of the existence of such processes—without some sort of socialization and control, no coherent "society" could be identified.[9] But any particular pattern implies winners and losers or a particular distribution of benefits and costs. It is worth asking how this distribution might shift if, for example, the view of energy as an ecological resource were to displace the commodity view. If the cost effectiveness question asked on the airplane is in some measure a product of socialization, who gains and who loses; if it were to be replaced by some other question as a result of some modification in the socialization process, how would the distribution of benefits and costs be altered?

The *effects* of conspiracy do not require that any explicit conspiracy exist. It may be that "What we see here instead is an ongoing social process in which scientific knowledge, technological invention, . . . corporate profit [and perhaps other forces] reinforce each other in deeply entrenched patterns, patterns that bear the unmistakable stamp of political and economic power."[10] *Effective conspiracy* in the sense of widely shared perspectives of established interests acting in society *as if there were* a conspiracy among those interests may well be worth considering. Even in their most innocuous manifestations, routine processes of socialization themselves amount to a kind of effective conspiracy to reinforce certain possibilities and outcomes, and block or deny others. Energy outcomes must inevitably be affected, for example, by the fact that utility executives with no direct incentive to encourage energy efficiency improvements long shared certain interests with private firms in the business of building large coal and nuclear power plants. Officials of the big three auto manufacturers, who may also serve as advisers to the Department of Energy and to the engineering schools whose graduates they employ, can not be expected to have an overwhelming interest in the development even of electric vehicles, to say nothing of more radical alternatives involving actual reduction in the use of personal vehicles. Corporate interests and economic efficiency are generally best assured, at least in the short run, if government and consumers behave "rationally"—that is, if they make choices on the basis of the same economic comparisons that corporations use in their decision making. Indeed, anyone with an interest in the nation's economy, has a certain stake in recovering full returns on the investment of social resources that has been made in existing patterns of life *before* entertaining suggestions for any substantial change in those patterns. And that stake is inevitably reflected in our educational system, in our research priorities, and in public decision making generally, in ways that may have the effects of conspiracy even in the absence of explicit collusion or abusive intent.[11]

If the cost effectiveness question itself is in some measure an artifact of socialization and control processes—for example, of an abdication of choice resulting from acquiescence to rapid technological change or of pressures to accept certain observations and attitudes and reject others—what does this say, again, about the wisdom of the choices that emerge? If social controls are involved, whether as an implicit or explicit outcome of particular coalitions of interests, or even if those controls emanate from a somehow disinterested, self-generating social auto-pilot,[12] are humanity's broadest interests best served by the results of those controls? Is the decision making process even "democratic" or are established interests conditioning our responses to the point that we are no longer able to act freely in the sense that true democracy requires?[13]

How, indeed, do we determine a wise course of action with respect to energy and the future? How do we determine what is appropriate either as individuals or in actions taken in the name of society as a whole? Are the traditional questions enough? Are they appropriate and do they in fact guarantee "optimal" outcomes? Or do we need to look more carefully, ask and answer a host of questions about the question, and explore broadly ranging territories of thought and experience not ordinarily thought to bear any relation to the choice of fuels used in the heating of our homes or animation of the production systems fundamental to modern Western patterns of life?

A Note on Expectations

If this chapter is doing its job, it should now have the reader at least mildly intrigued by the twist it begins to give to the old hat energy problem.

Be warned, however; this book will *not* be describing The Solution. To the contrary, much of the intent in trying to tease out an alternative definition of the problem is precisely to drop the matter squarely in the reader's lap unresolved. At best, this book will take the advice of Elting E. Morison and make only a "start from the particular case." ". . . In the beginning, [Morison suggests,] think small; don't go after the entire scheme of things head-on. Such a thought, [Morison continues] parenthetically, runs somewhat against the American grain. We tend to look for the big picture, the city set on a hill, the great society, the whole long line of the new frontier, a glory that will transfigure you and me—and if not soonest at least by tomorrow afternoon."[14]

At the slightest sign of hope it is somehow difficult in the modern world not to make a mental dash for the whole enchilada. Whether this habit is indicative of laziness or of a subtle (to me inappropriate) pessimism, it will not be indulged here. Unwary readers, even those intrigued by the twists just imparted to the definition of the problem, will be disappointed by the incompleteness of the solutions or, more accurately, the approach to solutions we come to in the end.

Our mental habit of expecting The Solution may reflect acceptance of, or a belief in, mass society. We seem to hope, almost breathlessly at times, that Tatum, or Washington, or the Democrats, or some other savior, will finally "get things figured out" and that we can all then fall in line as we finally implement The Solution. Yet we should be warned away, even if only through informal observation and recognition of the fact that people often have very different perceptions of the world we live in.[15]

As we proceed through the analysis of succeeding chapters in this book, I mean to preserve and even enhance hopes that, in spite of human differences, we may continue to learn from one another. I will continue also in the hope that we may remain united in a sustained respect for each other's views, and in our attempts to get as little as possible in each other's way when differences do arise. Retaining these hopes, I mean specifically to question the notion either that we must all adopt the same Solution to the energy problem or that any individual, party, or government could conceive or describe The Solution in the first place. Such notions may well be repressive by their very nature, aggravating rather than helping to alleviate our problems at the outset.

This book will ultimately arrive at the "particular case" of the home power movement in which growing numbers of participants are installing their own renewable electric power systems. On the basis of the reconfiguration of the energy problem that occupies the first portions of the book, the home power movement will be described as one that is strictly "irrational" by traditional standards, but eminently "reasonable" and understandable from a broadened perspective. In "explaining" (and at a personal level, supporting and applauding) this innovative behavior, however, we will *not* become involved in a litany of renewable energy success stories. I wish to avoid even the political leverage of such a "pep rally" approach to the energy problem.[16] Such an approach can be a useful aid to the enthusiasm sometimes necessary to innovative action, but it, like other hollow images of The Solution, ultimately implies at least a partial disengagement from the gore and the guts of the problem.

This is not to say that I believe that home power systems could never become attractive to the great mass of energy end users, even to those who now wish to think no further about their energy systems than the flick of a switch. (At this writing, Southern California Edison, one of the nation's larger electric utilities, is moving to offer home power systems as an option to its customers; who knows what new preferences may yet emerge from a new regime of implemented and observable technologies?) Nor in calling attention to particular departures from conventional wisdom, first in theory then in practice, do I mean to suggest encouragement for every harebrained idea. In attempting to open discussion a bit to consideration of unconventional alternatives, I do not mean to suggest that we abandon our best understanding of what is and is not physically and materially possible, of what may or may not be "sustainable" in the long run, or (certainly) of what is and is not desirable from a human perspective.

Merely taking the effort to develop an alternative theoretical and practical perspective carries with it an implicit element of "advocacy" (though perhaps no more so than acceptance of traditional perspectives). At a certain level, this book can, in fact, be read as personal advocacy and technical argument for "solutions" like home power. Taken in this way, however, the book probably is not worth very much. The more important object here will be to present an alternative view of the energy problem and eventually describe a new body of practice in the handling of that problem, *not* because that practice offers The Solution, but because it appears to reflect a unique engagement with what I argue is the real issue raised by energy choices— How do we wish to live in the world? It provides us with a new place to stand and observe the landscape. As a single, certainly limited illustration of the scope of technical and sociocultural possibility, it yet opens up to us the range of that very possibility and the content and significance of the question, How do we wish to live in the world? In the end here, the intent is not to offer The Solution nor to entertain or sustain any illusion that The Solution has been found, but to drop the matter of energy more squarely in the reader's lap where ultimately and inevitably it must lie, burden that it may be.

TWO

So Much Fuss over Energy?

The reader may yet ask if we are not on the verge of making a mountain out of a molehill. All this fuss over energy? Why not just let well enough alone? Surely this is a job best left to engineers and other experts whose business it is to work out solutions to technical problems of this sort. If energy supply is not a technical question, what is? Does it really make any sense to begin debating basic assumptions and reexamining well-established patterns of life just because of a few oil supply disruptions that happened back in the 1970s?

Perhaps not. In fact, for at least the period of a single lifetime, running out of energy supplies directly sufficient to sustain human life is unlikely, although continuing growth in energy use will be very difficult on the basis of fossil fuel supplies alone.[1] Even in the long run, one might imagine that sources of energy such as nuclear fission (including the breeder reactor) would surmount otherwise effective social resistance to be deployed in the event that other supplies were depleted to a degree that threatened survival.

Yet energy is of unique significance in society. A certain amount of "fuss" seems justified, as the next two sections will briefly suggest, even on the most immediate practical grounds. More subtly, it may be worth considering the possibility that established assumptions and patterns of life are long overdue for reexamination in any case. Could it be that "energy problems" are mere indicators with the appropriate "fuss" actually over the more fundamental issues raised by thinking about energy? Could energy "solutions," in fact, prove to be less a matter of technological breakthrough than serendipitous by-products of a more basic housecleaning? Most of this chapter will be devoted to exploring questions of this nature in an attempt to get at the more formidable subtle significance of energy in society.

Classical Concerns: A Thumbnail Sketch

We would be remiss if we did not at least pause to note that energy is of unique significance from the most immediate practical standpoint: it is an essential input to the production of all kinds of goods and services. In fact, when we come right down to it, energy is of the utmost practical significance in that no life is possible without it.[2]

Human beings have extended their earliest dependence on more-or-less direct solar inputs, to include the fossil and nuclear energy now utilized in modern production, transportation, and consumption systems. In the United States, per capita energy use in 1988 amounted to approximately 100 times actual food energy consumption (assuming 80 Q/yr national energy use and an average diet of 2500 kcal/day). In food production alone, "modern agriculture" entails the expenditure of roughly ten times as much energy (fossil fuels for tractors, energy for irrigation, artificial fertilizers, pesticides, etc.) as is contained in the food.[3] Although this kind of dependence can surely be reduced through improved design of the devices we use to convert and use energy, the process of redesign itself, as well as the subsequent replacement of existing equipment by more efficient equipment, will involve substantial energy inputs. In the limit, there is no substitute for energy in sustaining life, human or otherwise. Given this circumstance, a certain care in making energy decisions seems only prudent.

Before digging more deeply, however, it may be useful to begin with a thumbnail sketch of classical energy concerns as they have emerged since the OPEC (Organization of Petroleum Exporting Countries) oil embargo of 1973. These concerns were first focused on a newly emerging gap between oil supplies and demand for oil in global markets by the embargo itself and by the rapid inflation and other economic dislocations that followed from the embargo's more than twofold increase in (nominal) world oil prices.[4] Efforts to locate substitutes for foreign oil, however, also began to raise questions about energy supplies in general, especially as past exponential growth in energy consumption was extrapolated into the future and compared with existing prospects for increasing fossil fuel and other energy production. There was much early concern, as well, about dependence on imports, given the strategic importance of oil in military operations and war-related production, and about the long-term economic implications of huge balance of trade deficits associated with the purchase of foreign oil supplies. (Even as late as the early 1990s, more than half of this country's $60 to $100 billion trade deficit was due to energy imports alone.[5]) Environmental worries, particularly those relating to acid deposition, global warming, and radiation and radioactive waste disposal, also took on growing significance as energy concerns over the two decades following the embargo.

Domestic oil production in the United States peaked in 1970, failing to recover even with the burst of production from Alaskan discoveries added after that year. Only three years after this 1970 peak, the OPEC oil embargo confirmed a shift in the balance of power in setting world oil prices in favor of foreign (especially Middle Eastern) producers as against U.S. and Western producers and consumers. Although actual energy supplies were never reduced by more than a few percentage points either by the OPEC embargo or by the subsequent disruptions of the 1979 Iranian Revolution, the gasoline lines, price controls, natural gas shortages, and other dislocations that followed in their wake, quickly came to be seen as a full-blown "energy crisis."[6]

The acute focus of both popular and official attention on energy issues in the 1970s and early 1980s stimulated a rapid and continuing succession of large scale studies of the energy problem.[7] These initially placed substantial emphasis on new sources of supply and the technical challenges involved in rapid expansion of nuclear power, coal, and other alternatives to foreign oil.[8] More recently, emphasis has shifted toward increasing the efficiency of energy use as a growing consensus has been reached that such an approach is often much less expensive than new sources of supply. Vast potential for electricity savings through more efficient lighting systems and motor drives, and for saving other forms of energy through "super-insulated" home construction and improved window systems, through improvements in automobile mileage, and through cogeneration and other improvements in industrial practices have emerged.[9] And it has been suggested that something like a third[10] of our energy use (if not more) could cost-effectively be eliminated without significantly altering present patterns of life. With this shift in emphasis from supply to efficiency improvements, concerns have also shifted somewhat toward new sets of challenges in the effort to ease prospective supply-demand gaps. We have begun to worry about the slow turnover of automobile, housing, and other energy using stocks. We worry about the institutional momentum that has been built up in traditional building trades and practices, in the production systems now employed in turning out automobiles and refrigerators, and in the habits of mind of homeowners, business people, and design engineers that long essentially ignored energy as a desideratum.

As official projections of future energy use have gotten lower and lower,[11] there has been some stepping away also from early views that energy and economic growth were directly linked. Under the prevailing view at the time of the OPEC embargo, it would have been difficult to imagine continued healthy economic growth without comparably increasing rates of energy use. But the 1979 National Academy of Sciences CONAES study,[12] for example, already argued that the ratio of energy consumption to GNP could be reduced to as little as half its then current value over a period of thirty to forty years. Such a decoupling would be essential to long-term economic growth based on fossil fuels as projections for the lifetimes of conventional fossil fuel deposits continue to portray oil and gas essentially as "decades" fuels,[13] coal as perhaps a "centuries" fuel at current use rates,[14] unless more exotic sources (e.g., methane clathrates[15]) should become viable. There are, on the other hand, two long-run nuclear possibilities, one fusion, the other fission reactors including the breeder, either one of which could produce the equivalent of all present energy use for a thousand years or more.[16] Renewable (sun-driven) energy systems also have the technical potential to supply present energy use indefinitely; present electricity use in the U.S. could, for example, be produced from the kinds of photovoltaic (solar cell) systems now in production, if deployed to cover less than 0.37 percent of U.S. land area.[17]

A startling number of our most prominent environmental concerns are directly associated with energy conversion and use and are, in this sense, classical energy concerns themselves. Global warming, projected to amount to an average of 1.5 to 4.5 degrees Centigrade for a single doubling of preindustrial concentrations of carbon dioxide in the atmosphere, would most directly be an outcome of the fossil fuel combustion that produces the bulk of the carbon dioxide emissions.[18] Acid deposition, primarily a result of auto and electric power plant emissions, is also a direct outcome of present patterns of energy conversion and use and continues to be addressed through federal regulatory actions such as the sulfur oxide emissions limits of the 1990 amendments to the Clean Air Act.[19] A long list of other significant environmental concerns are, in essence, energy concerns as well. Radioactivity and radioactive waste disposal (as highlighted by the effects of the 1986 Chernobyl nuclear power plant disaster and the continuing stalemate over radioactive waste disposal in the United States), ocean and other oil spills (e.g., the Exxon Valdez), classical air pollution and urban smog (e.g., particulates, carbon monoxide, and ozone), strip mining damage to public lands, and other problems stem directly from energy conversion and use.[20]

The link between energy use and environmental concerns is so tight that most analysts have singled it out as one of the critical elements in our thinking about energy. As one has put the matter, "the growing magnitude of effluents from energy systems has led to saturation of the environment's capacity to absorb such effluents without disruption."[21]

Other classical energy concerns could also be added here. To include one more, for now, many analysts express equity concerns with respect to the allocation of energy among people. It may be noted, for example, that Western industrialized nations use many times as much energy as do less developed nations on a per capita basis. Declining energy resource availability also tends to create hardships at the level of necessities for poor people more than for the wealthy.[22]

At an immediately practical level, then, a number of significant classical concerns about energy are worthy of note. Dependence on foreign oil supplies, domestic and global declines in the availability of traditional fossil fuel resources, and the environmental, economic, strategic, and equity implications of present and possible future energy systems all give us good cause to think carefully.

Running Out or Running Short: Energy and Social Organization

Although we may well begin to worry about effectively "running out" of fuels like oil and gas in the sense that their availability is thought likely to decline markedly in the near future, there may be a more immediate source of worries, in the most practical sense, associated with "running short" of conventional fuels. Modern society relies, after all, on extremely complex systems of specialization and interdependence. The sustained production of food and of energy itself relies on the cooperative efforts of large numbers of people over extended geographic areas and periods of time. As Lewis Mumford has suggested,[23] such arrangements are uniquely vulnerable to disruption (i.e., "inherently unstable"), when compared with older, smaller scale systems of production "resting mainly on human skill and animal energy" in which each of many smaller groups made relatively "discreet use of the gifts of nature." When large scale "civilization" has broken down in the past, populations have been small enough and residual knowledge and social organization have remained coherent enough for a return to preindustrial modes of production generally sufficient to sustain food supplies. But the world's population before the beginning of the industrial revolution was only about one tenth to one fifth of what it is today.[24] It is difficult to imagine even today's population of about 5 billion managing to feed itself in the event of a real breakdown of present large scale cooperative production systems.

The real, short-term survival threat of energy shortages is not so much that energy supplies would be physically inadequate to feed even present populations, but that increasingly severe shortages or supply disruptions could introduce fatal instabilities in the complex systems of specialized interdependency that underlie present production systems.[25] This concern may be aggravated by the fact that the development and deployment of new modes of energy production (e.g., nuclear or solar) generally involve a significant period during which the new system constitutes a sort of energy sink rather than an added source of supply; during the period in which many new plants are being built to replace, for example, fossil fired systems, more energy would be invested in the construction of the new plants than they could put out (even if, over their lifetimes, the new plants may be substantial net energy producers). If traditional sources of energy are pushed to the point of exhaustion before a decision is made aggressively to pursue alternatives, the dynamics of the transition to new sources could still prove fatal to the complex social arrangements essential to accomplishing that transition. The global disruptions produced by reductions in oil production alone of a few percentage points in the 1970s offer one of the more visible indicators of the kinds of problems one might wish to invest careful thought in avoiding.[26]

The business of making energy choices is, in this sense, not quite like other entrepreneurial technology development efforts. If we fail in our attempts to develop high-definition television or fail to win in the commercial race to market this technology, we will certainly survive the penalties. If we drop out of the competition to develop supersonic passenger planes as, for some time at least we have, leaving the field to the British-French Concorde, the consequences could even prove beneficial to us, depending upon the still uncertain future of this technology. Failure to provide timely and adequate energy supplies affordably suited to the infrastructural needs of the time, on the other hand, could be catastrophic. The energy game can include uniquely drastic sanctions and, in this sense, is no "game" at all.

Energy Technologies as "Forms of Life"

Significant as these practical concerns may be, the real fuss over energy is probably justified more by subtler links with a formidable collection of social, political, and cultural issues that take us to the very roots of how we wish to live in the world.

When we begin thinking about energy futures, we soon must ask ourselves, for example, whether continuing population and economic growth are appropriate policies for this planet and, if so, for how long and at what rates? Energy concerns raise this sort of question in the sense that when we think about energy we need to know how many people we wish to supply and how much we can expect each of them to be using. Growth questions in turn raise concerns about the sustainability of present agricultural and other production processes from a broad range of resource, environmental, and social perspectives.[27] Resolution of these fundamental issues that would favor low or no growth would surely ease energy concerns both in the short run and long run. High growth rates raise bigger problems in terms of securing new energy supplies. But the implications of different growth choices for how we live are profound and pervasive.

As another example, concerns about our dependence on foreign oil, much of which is used to produce fuels for the private auto, can be taken as questions of demographics and of social organization with relation to our dependence on the auto itself. Would it make sense to reconsider historical trends and again locate homes, jobs, and appropriate shopping in such a way as to reduce auto dependence, even placing them again within walking distance of each other? Such a choice would surely ease the otherwise seemingly insoluble problems of pollution and congestion associated with the automobile while serendipitously reducing the need for fuels or for new technologies (e.g., electric vehicles) with the performance characteristics of the present automobile. Again, however, we are talking about some pretty basic changes in the way we do business.

In abstract terms, a society's choice of technology—and energy technology will prove no exception—does a great deal to shape or define the relationships it may encompass among people and between people and their natural environment. Technology might even be broadly defined as that which lies between a person and his or her human and natural environment. As Langdon Winner has eloquently put it, we do not simply "use" technology. We are as much shaped in the choice and use of technology as the technology itself may be.

> As they become woven into the texture of everyday existence, the devices, techniques, and systems [i.e., technologies broadly conceived] we adopt shed their tool-like qualities to become part of our very humanity. In an important sense we become the beings who work on assembly lines, who talk on telephones, who do our figuring on pocket calculators, who eat processed foods, who clean our homes with powerful chemicals. . . . In a trivial sense it is true, for example, that "You can always turn off your television set." But given how central television has become to the content of everyday life, how it has become the accustomed topic of conversation in workplaces, schools, and other social gatherings, it is apparent that television is a phenomenon that, in the larger sense, cannot be "turned off" at all. Deeply insinuated into people's perceptions, thoughts, and behavior, it has become an indelible part of modern culture.[28]

Its shaping effects may be all but invisible to us—we have no direct access to an alternative reality from which those effects might be more visible[29]—but the influence of technology choice in determining who and what we are is pervasive.

In its shaping of human relations, technology can be very "lawlike," separating, for example, racial or economic groups as in the case of the low bridges adopted in highway construction planned by New York's Robert Moses. (The low bridges permitted wealthier auto owners to commute from the suburbs and to get to and from the beaches, while preventing the passage of buses that were the only means of transportation available to the poor.[30]) Over time, as Langdon Winner suggests, the inherent politics of technologies like the atom bomb, as well as the political choices implicit in a host of specific design and development decisions, can take on the character of a political constitution, effectively answering the "age-old political questions [of] membership, power, authority, order, freedom, and justice."[31] Membership for decision making purposes, for example, may gradually shift from the original sense of citizenship to some notion of technical qualification, with power and authority newly defined in terms of particular forms of technical expertise. Notions of freedom may shift from the liberty to live as we choose, envisioned by some of the founders of this nation, toward the "freedom" of "producers" and "consumers" living within a modern industrial economy.

According to Winner, new communications technologies have, in fact, contributed to "an extraordinary centralization of social control in large business corporations, bureaucracies, and the military." Along with its undeniable benefits, technological "progress" has contributed to the emergence of an "undisguisedly authoritarian" workplace and to the "tendency of large, centralized, hierarchically arranged sociotechnical entities to crowd out and eliminate other varieties of human activity." Industrial techniques tend to eclipse craftwork, agribusiness makes small scale farming all but impossible, high-speed transportation crowds out slower means of getting about, and so on.[32]

In these and other respects, "technologies" must be regarded not simply as "tools we use" but as "forms of life."[33] Because energy technology, along with its associated energy sources, both "animates" virtually all technology and itself constitutes a major component of the whole of technology, it goes a long way toward defining who we are and how we live.

Energy and the Way We Live

If modern Western societies are distinguished from "primitive" cultural patterns by their technology, equally effective distinctions can be drawn in terms of the energy systems essential to the production, operation, and maintenance of that technology. If human relationships are shaped through communication by telephone, television, or print publication, then those relationships are shaped by the energy systems essential to the production, operation, and maintenance of those same communication systems. If human relationships with the natural world are affected by the use of automobiles, tractors, and spacecraft, then those relationships are also affected by the energy systems that are an integral part of such devices. And, with equal assurance, the relationships established in a society powered only by draft animals like horses will be different from those necessitated by a society that relies heavily on nuclear power production.

In fact, the very way we take up with the world may be a strong function of our historically unique use of energy. Empowered far beyond our physiological capabilities to alter what is around us, we are, "In our epoch, . . . always already oriented in the world as challenged forth by technology," according to one school of thought in the philosophy of technology. "Under the sway of [that] technology we understand the challenge of our age as one of getting everything under control as resources. Such an understanding and approach impoverishes what it touches: the farm field, the river, the forest, and even the planet itself." Empowered as we have been, there has been no barrier to a "petty homocentric" outlook, little opportunity to know what it would be to "let things be" or to develop a "bicentric" view in which all things, not just human beings, have significance and form unique connections with the rest of the world.[34]

It is perhaps no coincidence that energy production and use are closely linked with modern environmental problems. It is precisely our harnessing of vast fossil energy resources that has made it possible for us to behave like the proverbial bull in the china shop with respect to the environment. Concerns about global warming (stemming primarily from carbon dioxide emissions from fossil fuel combustion), acid deposition (associated mostly with power plant and automobile emissions from fuel combustion), conventional air pollution, nuclear waste disposal, oil spills, strip mining damage, and so on graphically illustrate the ways in which our choice of energy technologies directly shapes human relationships with the natural environment. Even problems such as stratospheric ozone depletion and the relationships with the natural environment it appears to imply must be regarded as a close function of energy choices. The link can be drawn in this case both in the sense that refrigeration and air conditioning (for which much of the culprit chlorofluorocarbons, or CFCs, are produced) only occur where certain energy supply choices have been made and in the sense that the large scale production of CFCs is possible only through certain processes of energy conversion and utilization. In a more general sense the colossal disruptions of natural ecosystems associated with modern agriculture, deforestation and other resource extraction, urbanization, and other human activities—even with present human populations alone—are difficult to imagine without the energy intensive technologies of the industrial era. Our very perceptions of time and space have surely been altered by energy intensive transportation systems that make us all potential global and even extraterrestrial travelers.

The energy-environment link is *not* a linear relationship; the very large energy disruptions of thermal emissions from power plants or cities are, for example, of relatively little, localized consequence environmentally, whereas the environmental impacts of ozone depletion from the much smaller energy conversions of CFC production are of greater and more pervasive concern. The link between energy use and environmental disruption is, nevertheless, close enough to raise real questions about the desirability of futures that would likely flow from "cheap and abundant" energy, whatever its source.

Human relationships, as well, are strongly influenced by our energy choices, perhaps primarily through the technological systems they support but directly in important ways as well. If one reflects on the elaborate social structures associated with an electric utility, for example—the educational traditions, the relations of technical expertise, the articulations of economic power in financing and of political power in regulation—some of this structuring of human relationships begins to be apparent. Experience has amply demonstrated that energy technologies such as a modern coal or nuclear power plant cannot simply be dropped into third world settings characterized by quite different patterns of social interaction. The power plant itself is inextricably linked to unique consumption configurations with all of their associated infrastructure (e.g., the manufacturer as well as the user of a toaster) as well as with the production side and its support structures (e.g., the utility transmission and distribution system, other generation technologies that must complement the coal or nuclear "base load" plant, and the lawyers, economists, engineers, managers, and stockholders that maintain and support the production system). A power plant, like many other energy technologies, is part of a "technological system."[35] And that system alone has pervasive implications for how people live and relate to each other.

One need hardly look beyond electric power to get a sense of energy technology as "technological system." An even more powerful illustration is, however, ready at hand in the system centered in petroleum and natural gas. From international relations and the Gulf War, through the gas station on every corner, to "cruising" as a ritual of adolescence, these systems almost seem to define the modern world. Oil and gas have transformed our common experience of the coldest winter day into mild summer, joined shores of the Atlantic and Pacific by a few hours plane ride, and transformed agriculture from animal husbandry and a refined sense of nature's nurture to an exercise in finance and equipment maintenance.[36] It is difficult to name any aspect of our material existence untouched by oil and gas.

Except for some of the more obvious of the environmental links just mentioned, the implications and constraints associated with particular energy technologies for social and individual life are rarely noted because they are ordinarily such an integral part of our ongoing lives as to be virtually transparent. Only in rare or extreme cases do we debate or even take notice of such effects. Only in the energy crisis era of the 1970s, for example, was there serious, widespread discussion of issues such as decentralization and the greater social diversity and personal freedom of choice some argued would be associated with so-called soft (renewable) energy technologies as contrasted with hard (largely coal and nuclear) energy technologies. As Amory Lovins put the matter,

> Perhaps the most profound difference between the soft and hard paths is their domestic sociopolitical impact. Both paths . . . entail significant social change. But the kinds of social change

needed for a hard path are apt to be much less pleasant, less plausible, less compatible with social diversity and personal freedom of choice, and less consistent with traditional values than are the social changes that could make a soft path work. . . .

In [a hard path] world, your lifeline comes not from an understandable neighborhood technology run by people you know who are at your own social level, but rather from an alien, remote, and perhaps humiliatingly uncontrollable technology run by a faraway, bureaucratized, technical elite who have probably never heard of you. Decisions about who shall have how much energy at what price also become centralized—a politically dangerous trend because it divides those who use energy from those who supply and regulate it.[37]

By contrast, Lovins associates soft path futures with lifestyles of "elegant frugality" and with traditional values such as "thrift, simplicity, diversity, neighborliness, humility and craftsmanship."[38] Although these characterizations of the human implications of hard and soft energy technologies have been subject to much debate, and particular characteristics would be shaped in part by the precise configuration of each that is deployed,[39] Lovins's work probably goes farther than anyone else's in bringing to our attention the extensive implications of energy choices for human interactions.

Occasional notice has also been given to some of the specific sociopolitical implications of particular energy choices, such as the potentially extreme effects on civil liberties that might be associated with adoption of plutonium recycling in a nuclear power cycle. In a classic article, Russell Ayres[40] argued that the associated hazards of plutonium diversion would lead to surveillance and other activities designed to prevent diversion, as well as draconian search, seizure, interrogation, and other activities aimed at recovery, should a diversion occur. These measures would amount to massive, though probably justified (and legally justifiable), invasions of civil liberties normally guaranteed by the Bill of Rights.

Permeating the very fabric of our lives, the effects of energy choices on the way we live are surprisingly difficult to discern, these few efforts notwithstanding. Chapter 6 in this book could be read as another round of efforts to sort out some of these permeating effects of energy choices as they have begun to be revealed in experimental settings.

In any case, all the "fuss" need not be simply about the practical necessity of assuring essential energy supplies. When we begin thinking about energy we soon find ourselves thinking also about how we grow our food, how we build our homes, how we transport things and people from place to place, how we produce goods and services in the economy and *what* goods and services we choose to produce. If we look at the issue long and hard enough, we begin to wonder how all of these decisions have been made, how it is that we build homes the way we do, for example, and how we have come to be as dependent as we now are on automobiles, clothes dryers, air conditioning, processed foods, and all of the other energy consuming technologies we have become accustomed to. We are led, almost inevitably, to consideration of how our society is arranged, how we make decisions, how we live our lives. This is where the fuss arises.

The "Technological Fix"

There will remain those, of course, with a simple faith that "technology" will save the day—that is that the world and human intelligence are constructed in such a way that we can depend on the development of new technical solutions to essentially any future problem including any possible energy shortage. This could prove to be true in either a limited sense or a strong sense. In the limited sense, energy shortages might prove resolvable while, for example, the environmental or civil liberties side effects of the "solutions" would simply have to be accepted as part of the package. In this limited form, the problem remains that the "solutions" might prove highly undesirable on many counts even if they did permit continued survival in the sense of continued production of food and other necessities.[41] A strong faith in technology might be justified if problems like energy shortages could be solved *and* their environmental or other side effects could through technical means also be avoided. To hold such a strong faith in technology, however, it would seem that one must either subscribe to an extreme faith in human ingenuity or conceive of nature as an almost infinitely pliable, even positively cooperative, situation for human life—or both. Even under the strong version of this faith, social decisions about energy perhaps should not be based on the simple expectation of saving breakthroughs without considering the possibility of objections to the social and personal exigencies implicit in the philosophy of "a break-through a day keeps the crisis at bay."[42] Do we, as in one very narrow example, wish to live in a society in which those responsible for generating the next round of technical solutions hold a status and exercise power commensurate with the absolutely essential nature of the service they would (attempt to) provide?

A more important question can be raised in defense of careful thought about energy choices, though, I believe. That question is, To what degree are "technological fixes" really desirable in the absence of a broader reexamination of our lives, of society, of natural constraints, and of our goals and objectives? To the degree that any problem or constraint we may encounter involves opportunies for learning and growth, a world in which technical fixes can be depended upon would be rather uninteresting, at least for anyone not directly involved in developing the fix itself. Although explicitly avoiding simple solutions to problems would perhaps be overdoing it, we might just stand to gain quite a bit by grappling with the energy problem in all its complex ramifications, ourselves.

In the article in which he coined the term "technological fix," Alvin Weinberg[43] specifically suggested that technological innovation and development be pursued as a simpler way of solving social problems than social approaches that would depend on efforts to persuade everyone "to behave more rationally." "Probably the most important new Technological Fix is the Intra-Uterine Device for birth control," he suggests. "Before the IUD was invented, birth control demanded very strong motivation of countless individuals." Similarly, the "brilliant advances in the technology of energy, of mass production, and of automation have created the affluent society," effectively resolving the age-old and apparently socially insoluble problem of maldistribution that once seemed to condemn many people to a life of poverty.

When his article was published in 1966, Weinberg's "technological fix" for possible future energy and other problems was, of course, nuclear power. As Weinberg put it, "The Technological Fix accepts man's intrinsic shortcomings and circumvents them or capitalizes on them for socially useful ends."

But Weinberg's endorsement of the technological fix suffers from at least two failings. It fails, first, to take into consideration the likely effects on people and on society of a policy of avoiding the work of grappling directly with real problems. What, other than the kind of mass society described in Bettelheim's words in the previous chapter, can we expect to emerge from such a policy? How can individuals or a society oriented around the technological fix avoid developing the character flaws of the spoiled child—no matter how good the intentions of those who strive to provide the technical fix itself? What sort of self-awareness or worldly wisdom can we expect to emerge from a society that implicitly seeks to avoid the lessons derived from the fact that certain things are impossible or that certain courses of action are inappropriate and that certain difficult choices must be made— lessons that every spoiled child, in fact, eventually encounters no matter how much the parent may wish to respond to every desire?

Weinberg's technological fix is also flawed in that it assumes that the only alternative in solving social problems is the less tractable process of social engineering.[44] Technological fixes are especially attractive for him as an alternative to the "traditional methods" whereby unnamed authorities attempt to resolve problems "by motivating or forcing people to behave more rationally." (In an energy context, one may be reminded here of efforts to "get people" to conserve energy.[45]) But the very process of developing technological fixes for society may well undermine the development in individuals of a capacity to see, independently, what sort of behavior might *be* more "rational" and in the process increase the scale of the problems to be dealt with by either of Weinberg's methods. More important, neither approach to the solution of social problems envisioned by Weinberg assigns any substantial, active role to ordinary people either in the definition of problems or in the development and implementation of solutions. The "engineered" population, whether socially engineered or the recipient of technological fixes, is apparently to be a virtual nonparticipant in the process of problem definition and resolution that is fundamentally constitutive of social and environmental relations.

Now a failure to engage the energy problem itself or the range of questions it raises may be expedient. It may even be that energy is not the proper focus for a broad reexamination of social and environmental relations but, instead, a more routine issue, properly resolved by routine methods. The flaws in a "technological fix" approach really become pernicious only if we adopt such an approach not simply with respect to energy issues but as a generalized policy. In its generalized form, the technology fix amounts to a kind of "delegation of life"[46] or of the interactive, decision making functions of life. In a very advanced form, under which even those responsible for developing the fixes, themselves cease to grapple with broad questions and only apply the methods of physical manipulation and the rules of cost effective substitution within a fixed pattern of life, the policy of the technological fix may become a kind of technological "somnambulance"[47] in which even the "fixers" display more the character of methodical automatons than of living, breathing, human beings.[48]

If we leave the energy question to technical experts, if we simply await a "cost effective" alternative to vanishing fossil fuels, if we do not actively engage newly apprehended constraints and the actual definition of important questions, if we do not actively consider possible changes in patterns of behavior consistent with our own carefully assessed values and objectives, and if we do these things as a matter of habit or routine, then we do, in effect, await a "technological fix." We cease to function as free and autonomous beings and we invite the atrophy of our own capacities for observation, reaction, decision, and renewed action.

All the "fuss," then, is not just about energy itself. Nor is it simply about energy as an expression of how we live our lives. We must now add that the fuss is, in a certain sense, over the degree to which we *do*, in fact, live our own lives rather than leave them to others to manage for us. All the fuss is aimed not just at the fueling of particular activities within a fixed pattern of life—important as this may ultimately prove to be—but at the multifaceted opportunity the network of energy issues affords for our own growth and development as individuals and as a society.

As a footnote to this discussion it may be worth asking why we seem to resist involvement in the broad questions raised by energy issues—if indeed we do. There is, of course, the natural tendency to conserve effort. If there is no need to become involved in complexities, why not save the effort for other more pressing matters? But is this all there is to it? In seeking to "refuel" present patterns of life without examining or adjusting those patterns even to the extent of paying higher prices for our fuels, are existing patterns of life really believed to be so close to optimality? Do we believe that *any* shift in the way we live that might ease energy difficulties would amount to a departure from optimality—that is, a concession and a net loss? Is it not possible, for example, that smaller cars, bicycling to and from work, or shifting work places closer to homes as a response to transportation fuel pressures, might be found beneficial on balance (e.g., easing pollution, congestion, or other difficulties) in their own right? Have we simply "arrived?" Is there really nothing but a "holding pattern" beyond this? Is there no point in even looking for opportunities for further development?

Or do we, on the other hand, simply lack the time to examine the way we live—and, if so, is this in itself not an indictment of the way we live and an urgent sign that our patterns of life are in need of reexamination and adjustment? Is it possible, even, that there are elements of our way of life that we firmly *know*, at some level, to be inappropriate or deeply dissatisfying, but that we would prefer not to face directly? Are we, for instance, simply trying to enjoy the materially easy life of energy abundance, not anxious (though we know it cannot last) to curtail the ride and confront the constraints and disciplines of life beyond the amusement park?

THREE

Surely the Experts Have Thought the Matter Out

In an age in which we know so much, there is a tendency to think that there surely must be a "scientific" answer to the energy question. We must, after all, be practical about so important a problem. It should be addressed rationally. Even if we grant that the decisions should ultimately be arrived at democratically, should we not be clear that they do not rest on purely romantic notions, that they are not merely products of the politics of uninformed special interests?

Surely the government or policy makers or experts somewhere have thought this matter out carefully—after all, is that not their job? Is that not why we regard them as our leaders? Is that not why government officials, university professors, heads of large corporations, and others are highly rewarded financially and in terms of status? Surely these social leaders are the ones with both the time and the tools needed to sort out what should be done about our energy problems both directly and, where appropriate, in their complex ramifications.

In fact, two interrelated expert responses to energy issues have remained prominent since before the arrival of the energy crisis in the early 1970s. With very few exceptions,[1] these well-established responses have framed the major publicly funded energy studies,[2] extending back well before energy became a major public issue,[3] and they continue to provide the rationale for the full range of U.S. energy policies.

On close examination, however, both prove to be less robust and less conscientiously constructed than might be hoped and expected. Both prove far less complete as justifications for traditional policies than might widely be assumed.

The first, or "engineering," response consists essentially of the effort on the part of scientists and engineers to come up with a technical fix—that is, to develop fusion energy systems, synthetic fuels from coal or oil shale, safe and environmentally benign fission and breeder reactors, much less expensive versions of renewable energy systems like photovoltaic electricity, or other substitutes for conventional energy sources. The engineering response generally takes society and people's goals and objectives as known and invariant. And it accepts, for the most part, the second expert response centered in the economic view introduced in Chapter 1, under which the appropriate measure of the feasibility of particular technical options is economic. Although the second, economic, response is open to a variety of ad hoc adjustments for environmental and other factors, it ostensibly relies on objective measures of preference (i.e., market behavior) and on quantitative cost comparisons to select the "best" (most "cost effective") energy options. It at least implicitly rejects as irrational any choice of energy systems more expensive than other alternatives unless dollar values can be assigned to costs or benefits not accounted for in the marketplace that will make the chosen systems actually cheaper than the alternatives. Although the economic perspective adds certain constraints to the engineering approach, the two responses together amount in practice to the search for a technical fix.

Characterized in this way, it would be surprising if both of these expert response modes did not suffer from the failings of the technological fix approach as outlined in the previous chapter. Applied in their strictest form without ad hoc political intervention, they place energy choices with all of their social and human consequences on a virtual autopilot, transferring many of the attributes of life (to paraphrase Mumford[4]) to what amounts to a choice making machine. Even as they are applied in practice, they generate only hollow echoes of past patterns and assume away most of the human concerns raised by the energy problem.

Let us briefly illustrate the kinds of problems that arise under each of these expert responses to energy problems. First, how does the engineering expert response assume away most of the human elements of the energy problem?

The Engineering Response

To begin with, the engineering response tends to assume that human values or value-based behaviors are known, temporally invariant, and globally uniform. This assumption is clearly evident, for example, in the prestigious and much publicized National Academy of Sciences CONAES (Committee on Nuclear and Alternative Energy Systems) study, *Energy in Transition: 1985–2010.*[5] In this and other studies in which an engineering response predominates, it is assumed that human values are such that energy supplies will ultimately be needed to support Westernized patterns of production and consumption everywhere in the world: "the industrialization that is an inescapable aspect of economic development will greatly increase [third world] reliance on electric power, of which they now have very little. Their agriculture will also shift from animal and human energy to tractors, harvesters, and trucks, and from natural to industrial fertilizers. As personal incomes rise in these countries, they will want better housing with more lighting and appliances, not to mention air conditioning. The more affluent of their citizens will demand motorcycles, automobiles, and air travel."[6] Even if we grant that the people of third world nations might ultimately prefer to live at something more than a bare subsistence level, are we justified in assuming that everyone wants to live "just like us"? Given the social movements of the 1960s and 1970s alone, do we really know with such certainty what people want, even today, in our own communities? Can our questions be answered by expert responses that begin, as Nader and Beckerman[7] have charged, with the assumption that *everyone* accepts the "lifestyle ideals of the upward mobile middle class?"

Even if we were to assume that environmental and other conditions were to remain entirely unaffected by future energy use and that conventional fuels would continue to be available at current or historical prices, can we really say with such assurance what the peoples of our own nation, let alone the remainder of the world, will want between now and the year 2010? It would seem that one of the profound and unresolved questions raised by new constraints on the supply and use of energy would be precisely this: what preferences will emerge and predominate as human values operate on the new regime of technical possibility now taking shape? Simply assuming that we know present desires, that these will remain invariant over time, and that everyone in the world must subscribe to these same values, plainly begs one of the most interesting and important (i.e., one of the determining) questions to be answered here.

In a more specific instance, what may be characterized as an engineering expert response has in the past contributed to a "do nothing" policy on the issue of global warming that might occur largely as a result of continued burning of fossil fuels. A 1983 study conducted by the National Research Council, for example, concluded that "the temperature increases of a couple of degrees or so projected for the next century are not only large in historical terms but also carry our planet into largely unknown territory."[8] Assuming, however, that the likelihood of coordinated international efforts to curtail fossil fuel combustion and resulting carbon dioxide emissions is "negligible," the study authors' central policy recommendation was simply to "anticipate changing climates."[9] Here again, the critical questions raised as a part of the energy problem—such as what individual and cultural responses to new energy-related conditions are possible and what responses might emerge or might be preferred—have effectively been sidestepped by assuming that the new situation will simply produce nothing new in the way of response. Energy problems specifically challenge us to weigh long-term climate risks against cheaper and more convenient energy use patterns in the shorter term. The expert response in this case is at best a default in favor of cheap and convenient energy use and at worst a kind of nihilist argument that there is no point in any intellectual endeavor to exercise judgment in the matter.

The engineering expert response also tends to take existing institutional arrangements as invariant or to give those arrangements priority over competing thrusts for change.[10] "A problem for many solar energy alternatives is finding ways to introduce decentralized technology into a centralized network without disrupting the economics and reliability of the network."[11] This is, of course, one way to look at the matter. But do we necessarily wish to give the existing institutional framework (i.e., the centralized utility network) priority over the new technology (i.e., solar energy alternatives)? This is a question the engineering response characteristically fails to note or raise. But does this priority necessarily best serve social interests? Or is it possible that the existing institutional framework, despite its greater size and longer history, should be reshaped to meet the needs of the new technology? Here again, the engineering expert response to a large degree begs—even prejudges—the fundamental questions with respect to desirable energy futures. Should we be aggressively pursuing a transition to renewable energy futures, or should we be making every effort to shore up important elements of the existing framework for energy delivery? There are plenty of historical instances (the automobile would be one classic example) in which new technologies have powerfully disrupted and even displaced previously well-established social institutions; how were these situations different and why should we only begin our thinking with the assumption that they are different?

As a further illustration of difficulties with the engineering expert approach, it tends to assume that human considerations must give way to technical considerations. Berkeley's Professor John Holdren, in fact, specifically challenged this practice as evidenced in the CONAES study by adding the following dissenting footnote to the CONAES report:

> The [study report's] assertion that coal and nuclear fission are the "only readily available domestic energy sources that could even in principle reverse the decline in domestic energy production over the next three decades" rests on a judgment I do not

share. This judgment is that the obstacles to significant penetration of the energy mix by renewable energy sources in this period are more fundamental and less tractable than the obstacles in the way of expanded use of coal and nuclear fission. The obstacles for the renewables are *technical and economic*—extensive penetration between 1990 and 2010 would require some technical breakthroughs yielding large cost reductions early in the period, or willingness to spend significantly more for renewable energy supplies than we have been spending for conventional ones. The obstacles hindering coal and nuclear are different— they are *environmental and sociopolitical* more than technical and economic—*but they are neither less real nor more easily circumvented than the liabilities of the renewables.*[12]

Here, again, the engineering response begs the fundamental questions. Would it be in our best interest to accept the environmental and sociopolitical costs of coal and nuclear alternatives or to accept the technical and economic difficulties and costs of renewable alternatives? The only answer to this question provided by the expert authors of the CONAES and similar studies is the answer implicit in their assumptions.

To the degree that scientists and engineers evenhandedly probe the realm of physical possibility, they serve an important, perhaps an essential social function. To the degree that they allow implicit or unexamined assumptions about people's values, the sanctity of particular institutions, or other human factors to govern their investigations, however, they risk the foreclosure of important avenues of social evolution and a substantial undermining of democratic decision making.

It must be recognized that the engineering expert response does not offer anything approaching a "scientific" resolution of the energy problem broadly conceived. There is no clear evidence that expert pronouncements from this perspective even recognize the far–reaching questions they assume away. There is no evidence, certainly, that superior knowledge or superior methods have been brought to bear in resolving such questions.

The Economic Response

Economic expert responses, the second of the two most prominent expert perspectives, prove to be similarly limited. In the first place, economic attempts to compare alternative energy futures depend heavily on prices. We often take market prices to be as much matters of fact as is the hardness or tensile strength of a material, but they are actually as much or more the product of agreements between buyers and sellers heavily conditioned by social convention as they are indicators of natural or physical conditions.

In comparing, for example, continued use of oil with some other energy source, what price should be assigned to oil? If it is priced on the basis of its historical or present production costs, its price should be zero or very low (certainly a small fraction of its current world price). If it is to be priced, on the other hand, at its "replacement cost," its price should be much higher, perhaps even infinite. Its production costs will also be affected by whether we consider only private costs or include also any social costs associated with oil production and use. If the social costs associated with the environmental effects of oil production and use were, for example, included in oil prices, those prices would assuredly be substantially higher than they are under the traditional practice of "externalizing" those costs from the oil transaction. On the demand side of the bargain establishing price, we must also consider the matter of how the demand population is to be defined. If it includes only those alive today who have the money to buy oil, then the price agreed upon will be much lower than it would be if everyone alive today were sufficiently wealthy to bid in the market for oil. If we expanded the demanding population to include all future generations that might have an interest in purchasing some oil, prices might be bid up even higher yet, well above present prices. "The point is that resource prices, faithfully determined by supply and demand, could range from zero to infinity, depending on essentially arbitrary conventional definitions of cost (historical versus replacement; private versus social) and on an arbitrary definition of the demanding population (present only or some number of future generations)."[13] Taking these and other complications fully into account, it seems that our energy decisions—that is, decisions about appropriate rates of resource consumption, about our willingness to accept certain forms of environmental damage, about how the interests of particular groups are and are not to be represented in decision making, and so forth—are more accurately price *determining* than they are price *determined.*[14] If prices are this "soft," we cannot assume that economic comparisons of energy alternatives will provide sound guidance unless we first examine carefully all of the social conventions incorporated in market pricing and determine that each of these is and will remain firmly justified. Yet economic analysis of energy options and the reporting of results and recommendations occurs routinely without even taking note of these complications. Without the unique and courageous efforts of historically isolated economists like Herman Daly (quoted previously), there would be virtually no indication to the layperson that such a range of complications might exist.

As a second, pivotal difficulty with the economic expert response, economic comparisons of alternative energy futures or, for that matter, of virtually any other action alternatives, incorporate a critically important "discounting" assumption. This is the assumption that the value of a dollar you or society as a whole will hold at some point in the future is *less* than the value of a dollar held now. This assumption, which is made quite apart from the inflation issue, must ultimately be linked to an *assumption of continuing economic growth*: if it is true that putting a dollar in the bank or investing it in some other manner will, on average, return that dollar plus some interest at some point in the future, then the "discounting" assumption makes some sense. If, in other words, it is generally possible to trade say, $1.00 now for a guaranteed $1.06 a year from now, then it makes some sense to say that one dollar now is equivalent in value to a dollar and six cents a year from now. In this case, the interest rate would be six percent, and the reverse process for finding the present value of a future sum would employ a "discount rate" of six percent.[15] If growth of this sort is not guaranteed, the discounting assumption is difficult or impossible to justify.

The implications of the discounting assumption are far–reaching. If, for example, we make the unexceptional assumption of a 10 percent discount rate, the present value of the future sum of $100 billion is only $61,815 if that future sum is to be exchanged 150 years from now. If, in other words, we invested $61,815 at a 10 percent interest rate and waited 150 years, we would have $100 billion in our account. Under the discounting assumption with this discount rate it would therefore be considered irrational to spend more than $61,815 now to avoid anything less than $100 billion in damages that might accrue 150 years from now as a result of air pollution or other effects of energy use. If we knew, for example, that reduced fossil fuel combustion today would ease global warming and avoid sea level rise or other damages amounting to $100 billion 150 years from now, economic analysis based on the discounting assumption would tell us that we should not spend more than $61,815 today to avoid those damages. If we were to adopt a 20 percent discount rate, the present value of the same $100 billion would be only $0.13. More generally, the higher the discount rate is and the longer the time period in question, the more exaggerated the difference will be between the future amount and its present value.

The discounting assumption amounts to a systematic bias in energy, environmental, and other decision making in favor of courses of action that involve short term gains and longer term costs. Because discounting "shrinks" sums in the remote future more than those in the near future (just as more interest accrues on amounts left in the bank longer), any project alternative that can put the benefits in the near term and the costs in the more remote future will look relatively more attractive than a project with equivalent costs and benefits distributed with the costs first and the benefits later. An energy production system that may have very high long-term waste disposal, decomissioning, or environmental costs may yet be preferred under the discounting assumption to other systems that require high initial investments and provide energy benefits only over a lengthy operating lifetime. In these ways, the discounting assumption can profoundly shape both choices among energy alternatives and the weighing of environmental and other long term characteristics of particular energy futures.

The difficulty, again, is that the uniformly adopted practice of discounting simply assumes away some of the most basic questions raised by the energy problem. How do we wish to balance the costs and benefits our actions impose on future generations against the costs and benefits our actions impose on us? Is it appropriate, to take one extreme example, to "discount" the 100 to 800 deaths per nuclear power plant, projected to result on average over the indefinite future, at "reasonable rates, such as 5 percent," and express them in present value equivalents of 0.07 to 0.3 fatalities per plant year—that is, equivalent to roughly 10 fatalities per plant?

If in fact we regard an indefinite continuation of economic growth as desirable, is it *possible* on the basis of the energy programs ordinarily analyzed or is a substantial rethinking of the relationship between energy production and economic activity needed? Is the selection of preferred energy alternatives on the basis of "reasonable" discount rates even likely to maximize future economic growth if this is our objective, and over what period would growth be maximized?

It is, to be sure, difficult to be much reassured regarding possibly critical energy shortages or other severe energy-related social dislocations, on the basis of economic analysis that in its foundations *assumes* continuing healthy economic growth.

How do we know that decisions made on the basis of market prices, with or without the discounting assumption, will not lead us into catastrophic energy shortages at some point in the future? Surely this is a question economic experts have thoroughly examined and resolved before applying their analysis to energy questions.

Unfortunately, the reasoning here again may well prove circular. There is indeed a large literature on resource scarcity, much of it based on the idea that resource producers would, under steady demand conditions, tend to raise prices for their resources as those resources become scarce. Even in the absence of "market imperfections" like price collusion among resource producers, resource scarcity should, according to the usual argument, be signaled by higher prices. Critical resource scarcity is not a problem, it has been widely argued, because real prices (i.e., prices with inflation effects removed) have not risen appreciably over the long run even for oil. However, "The theoretical arguments of the conceptual and empirical literature on economic indicators of long run resource scarcity are logically flawed. If resource allocators were informed of the nature of resource scarcity, their behavior and the economic indicators it generates would reflect the scarcity. But if they were so informed, we could simply ask them if resources were scarce. If they are not informed, their behavior and economic indicators are as likely to indicate their ignorance as the reality. Unfortunately, there is no way of knowing whether they are informed or not unless we already know whether resources are scarce."[17] In his article elucidating the weakness of economic indicators of resource scarcity, Professor Richard B. Norgaard intimates that his discipline's failure to appreciate fully the implications of this logical fallacy would make a fascinating study in the sociology of science. He concludes his article, moreover, by suggesting that the classical treatment of resource scarcity indicators may simply be *un*scientific: "two beliefs about science have not significantly changed [over the years]: (1) science feeds on the tension between theory and reality, and (2) individual scientific arguments must be logical. The attempt to use economic indicators to determine whether resources are scarce over the long run has not met either of these criteria of what makes an endeavor scientific."[18]

There is, additionally, an elaborate mathematical theory purporting to show that allowing prices to take their natural course in a properly functioning market and making decisions based on those prices will naturally lead to socially optimal outcomes.[19] This "optimal depletion" theory, however, rests on a collection of assumptions that again beg the important questions. The assumption of a "properly functioning market" itself requires, for example, that market prices accurately reflect full social costs, not simply private production or other costs. More surprising, however, the theory implicitly assumes that no critical shortage of energy will ever occur (i.e., that prices will never become infinite[20]). It assumes that unspecified "backstop" fuels exist that can be substituted for energy sources that are depleted and that these can be developed in socially acceptable ways at socially affordable prices.[21] Optimal depletion theory effectively assumes that environmental and other externalities are nonexistent or inconsequential and that no socially preferred path would involve significant departures from recent patterns. (These last assumptions relate to the issue of "nonconvexities" and possible "corner solutions" in the mathematics of the theory and are made largely so that the mathematics will work out.[22])

If all of these assumptions had been demonstrated conclusively to be true, it would seem that we would indeed have little to worry about. The problem is that they are only assumptions. The extensive economic literatures barely begin to probe alternative assumptions and their possible implications, and fail to display any carefully constructed argument supportive of the assumptions adopted. While they provide highly refined answers to highly specialized questions, they effectively overlook, evade, or bury the determining issues.

The Failure of Traditional Methods

One of the most disturbing things about the engineering and economic expert responses is that they both tend to excuse or eliminate ordinary people from direct participation in energy-related decision making. As noted earlier, the engineering response makes very rudimentary assumptions about people's values, goals, and objectives, primarily shouldering the technical task of providing the means to those objectives. The engineering view thrives on an image of the public as *uninformed* and ill-equipped to make the only choices of real interest— that is choices among complex technological alternatives. Hence it is with no apparent concern that the CONAES chairman, for example, notes that a series of public hearings conducted in five major cities across the nation as a part of the CONAES study had resulted in no summary, "nor did they prove particularly fruitful."[23] The human questions, seemingly, have obvious answers while the technical questions are beyond the grasp of an untrained public. The economic view, on the other hand, takes people's preferences as the pivotal issue. But it assumes that those preferences are adequately revealed by market behavior. It assumes that resource producers and resource consumers are *well informed* and sophisticated in their energy-related choices so that prices and other market data are a sufficient basis for public policy making. There is no need for direct public participation except in the marketplace through which preferences are revealed.

The two responses make almost contradictory assumptions about the public: incompetent and poorly informed vs. highly capable and well informed. And they make contradictory (or complementary) assumptions about technology and human preferences: in one case the first is problematic and the second is obvious; in the other case, technology will take care of itself and we must attend carefully to the data on preferences. Yet each approach appears to offer answers to energy questions with no need for direct involvement on the part of the public in the energy issue. Beyond their failure to engage and directly grapple with the determining questions, both of these expert responses manage to place both the individual and society in an essentially passive role, as if their exercise of judgment and choice were entirely unnecessary in the resolution of the matter. The resulting risk of poorly conceived energy policies, and the long-term hazard of atrophy in the public's capacity to choose, have already been alluded to as failings of the "technological fix" approach.

Established engineering and economic expert responses offer little evidence of careful consideration of the broad implications of individual and collective energy choices. They certainly give no indication of having dealt definitively with human values and preferences, technological and institutional priorities, tradeoffs between the present and future, the potential for severe social dislocations related to the depletion of energy resources, or other pivotal issues. If the leaders involved have thought long and hard, they offer no convincing demonstration, by superior methods or otherwise, that their implicit or assumed positions should be accepted.

Additionally, *both* responses appear antithetical to the traditional *participatory* foundations of democracy: government "of the people, by the people, and for the people."

In their defense, it should be emphasized that reputable engineers and economists will be among the first to admit that their methods do not begin to offer complete answers to society's energy questions. Reputable engineers and economists offer the results of their research only as inputs into a much larger process of individual and social decision making and have no desire to eclipse or exclude a wide variety of other inputs into that larger process.

The problem, of course, is that the basic questions raised by energy concerns cannot be answered by any expert or scientific method. Although they can and, one would think, should be informed by knowledge of physical and social conditions, they can be answered only through the application of individual and collective judgment based on the values, preferences, and assessments of the individuals and societies involved. It may be that the expert responses already described have become so influential—unjustifiably influential—partly because they help to fill an apparent vacuum.[24] They act as substitutes for the sure-fire methods that do not and (I would sincerely hope) *cannot* exist.

Herman Daly[25] has put the matter in the form of a distinction between the active process of planning and expert methods based on prediction. It is appropriate, he writes, to "plan events subject to our control and predict events that are beyond our control." But the two are confused in dealing with energy issues. We could "make a collective social decision regarding energy use and attempt to plan or shape the future under the guidance of moral will," but we instead tend to treat that use "as a problem in predicting other peoples' aggregate behavior and seek to outguess a mechanistically predetermined future."

> Statisticians and econometricians frown on moral exhortation as unprofessional. Prediction sounds scientific and value free, while planning sounds value loaded and downright socialistic. . . .
>
> But, alas, prediction has in practice become implicit planning. . . . Outguessing a predetermined future and then treating the prediction as if it were determined by natural laws to which we must conform is what Robert K. Merton called "self-fulfilling prophecy"—or the elevation of trend to destiny, in Rene Dubos' phrase.[26]

This approach is simply "an unsuccessful retreat from the responsibility of choosing goals, and is unworthy of any organism with a central nervous system, much less a cerebral cortex."[27]

Close examination suggests, in any case, that attempting to answer our energy questions strictly on the basis of established expert responses is to mislead or to be misled.

FOUR

With Troubles Enough, Experts Differ

There are, of course, the dissenters—those who teach at universities, work at major laboratories, or have otherwise curried the markings of expert status, but do not accept traditionally established thinking on energy matters. We have already relied on some of them, in fact, in calling traditional thinking into question in the previous chapter. (There may also be "ordinary" dissenters—dissenters without any of the traditional indicators of expert status—but this is a category to pursue in later discussions.)

Three groups of dissenter arguments will be reviewed in this chapter. First, there is the group of physical and natural scientists as well as a few economists that argues that established economic responses fail to appreciate the unique and essential physical role of energy in the production process. A second group, perhaps best represented by the work of Herman Daly, suggests on the basis of environmental and other concerns that traditional economic objectives need to be deemphasized or reformulated in favor of efforts to make human patterns of life more sustainable. Finally, although they have rarely been brought directly to bear on energy concerns, the perspectives of historians, anthropologists, and other disciplinary "experts" can be taken to directly challenge both the engineering and economic responses as unrealistic and misguided.

The Physical Role of Energy in the Economy

Those who regard energy resources as ordinary commodities and resource depletion as a relatively minor problem often refer to the fact that expenditures for primary energy production represent only on the order of 5 percent of the Gross National Product (GNP).[1] In the 1970s and 1980s, they have also begun to buttress their argument for a modest role for energy in the economy by noting the decline of earlier concerns that continued GNP growth might be impossible without continued growth in energy production and consumption. The established conventional wisdom now holds that GNP and energy growth can be, or indeed have already been, decoupled; significant growth in GNP has occurred since the 1973 oil embargo although energy use over the same period grew relatively little.[2]

Dissenters, on the other hand, have argued that economic treatments of energy fail to capture the unique physical importance of energy in economic activities.[3] "In the average economist's view, the economy is a closed, self-regulating, self-driven cycle of goods and services flowing between 'households' and 'firms.' Although this simplifying assumption has allowed economists to describe and predict certain kinds [of] economic exchanges, it is unrealistic: the economy is not closed but requires inputs of resources, especially energy, from outside the economic system."[4] These dissenters argue that conventional economic production factors (i.e., capital, labor, natural resources, and government services) are, themselves, largely products of energy inputs and that, in this sense, energy is a much more important factor in the economy than is traditionally recognized. "Most standard models of production consider fuel and other natural resources to be qualitatively no different from other factors of production. As a result, many believe that natural resource inputs to production are 'small potatoes compared to labor, or even to capital,' and that 'reproducible capital is a near perfect substitute for land and other exhaustible resources.' . . . This view is inaccurate because free energy is required to upgrade and maintain all organized structures, including capital and laborers, against the ravages of entropy. It ignores the physical interdependence of capital, labor, and natural resources."[5] To take a concrete example, the energy required to build a car is not simply the amount of energy used in the manufacturing process itself, but includes also a portion of the energy used in producing the machinery used in the factory (i.e., in creating capital), in extracting and processing the materials incorporated in the car (i.e., in the utilization of natural resources), and in heating the homes and providing the food and other commodities that constitute compensation for the labor force (i.e., in providing labor as an input). Because none of these production factors would be available without their associated energy inputs, energy itself is seen to be of unique importance in the production process and therefore in the economy as a whole.

Robert Costanza[6] has argued on the basis of input-output analysis that, with the exception of primary energy production itself, the correlation between the total amount of energy "embodied" directly and indirectly in a product and its market value is very strong throughout the economy. Arguments of this sort tend to support an energy theory of value, although such theories are "summarily dismissed by neoclassical economists,"[7] and imply a much more important role for energy in the economy than energy costs as a fraction of total GNP might suggest. If, in fact, energy is the primary input to the economy, and other production factors (capital, materials, labor) are themselves in large measure products of energy inputs, it becomes much more difficult to imagine sustaining customary material standards of living if energy inputs were to be substantially reduced.

Costanza and others[8] also point to the fact that a large component of the increase in labor productivity over the past seventy years has been directly associated with increases in the amount of energy each worker has been able to apply to his or her task, both directly and through the energy embodied in industrial capital equipment and new technology. They argue that much of the decline in the ratio of energy use to GNP that is evident between 1929 and 1981 has been a product of changes in the fuel mix and shifts in the proportion of energy used in "final demand." Increasing proportions of higher quality fuels like oil and "primary" electricity from hydro and nuclear sources have increased the amount of useful work that could be done, and hence the economic output achievable, per unit of energy input. Similarly, reductions in the fraction of energy use going to final demands, such as the energy consumed directly in homes, have increased the economic output per unit of energy flowing into the economy. (Fuels burned in heating a home, for example, do not produce as large a contribution to GNP as the same energy embodied in the construction of a new manufacturing machine). "Eighty-eight percent of the decline in the E/GNP [energy/GNP] ratio since 1973 can be explained by the declining proportion of GNP spent on fuel by households."[9]

On the basis of such arguments, these dissenters (in the company of certain much earlier analysts[10]) suggest that we should give careful attention to the "energy return on investment," or EROI, in making our selections among energy alternatives. Energy return on investment, defined as the ratio of the amount of energy extracted from a given process to the direct and indirect energy consumed in the extraction and delivery process, may even be taken to be a more critical factor than market price in choosing among energy alternatives. Our interest in any energy system lies, after all, in the amount of energy it can supply us for applications beyond simply maintaining and replacing the system itself.

From such an alternative perspective it may be important to question the usual policy of waiting for resource price paths to cross (i.e., for renewables to become cheaper as oil and conventional fuel prices rise). Traditional economic comparisons arguing against the early adoption of renewable energy alternatives often seem to reassure those who believe that an eventual shift to renewables is inevitable by alluding to the common assumption that conventional energy prices will rise as conventional resources become more scarce and that we need only wait a few more years before the prices for longer term alternatives will compete with those for conventional fuels. If material production, including the production of renewable energy supply systems is, however, a strong function of energy inputs, can we really be assured that alternative energy prices will not simply rise with (i.e., as functions of) conventional fuel prices? Can we even be sure that conventional energy prices will rise in real terms as resources become more scarce: the closer the link is between energy inputs and material production, the more closely the price and availability of the one will be linked to the price and availability of the other. Declines in energy availability could simply lead to declines in material production generally, with no necessary shift in the value of one relative to the other.[11]

This first group of dissenters suggests, in fact, that recent, current, and future economic difficulties may well be ascribed in large part to changes in energy resource availability.

> underlying the energy crisis and the ensuing economic malaise was the declining physical availability of high-EROI petroleum, and a reliance on economic and political models that did not account for it.[12]

> it is important for everyone to understand that many of our economic problems stem from a deteriorating supply of fuels and other resources. . . . Society needs to adjust its expectations and strategies.[13]

Noting that there are no energy alternatives on the horizon with EROIs comparable to those of an earlier period of oil extraction, this group is unwilling to accept the assumptions of the economic expert response. "[P]ast experience with capital-intensive ventures such as fission and synfuels suggests that it would be unwise to assume a priori that fusion or any other proposed fuel source will necessarily have a large EROI. Although we should research aggressively all potential fuel technologies . . . we should also plan for the contingency that new high-EROI sources might not be found."[14] A dissenting view of this sort implies very different constraints and priorities for society than does the established engineering-economic perspective. Given declining conventional resource availability, it would be logically difficult to support recent energy policies calling only for modest adjustments in the usual patterns of energy production and use without also endorsing the established view that energy and the economy can be fundamentally decoupled. The belief that energy and material production are more profoundly intertwined would seem to call for a more extensive reexamination of energy practice and of the patterns of life that practice supports.

The Appropriateness of Economic Measures and Objectives

The second group of dissenting views introduced at the start of this chapter is closely linked with environmental concerns but also bears directly on energy issues. One of the central concerns expressed here is the idea that traditional economic objectives, like efficiency and economic growth, and traditional measures of progress, like growth in GNP, fail to capture environmental damages and other critically important desiderata as they shape individual and social policies. In our high regard for Adam Smith's "invisible hand," we fail to take into account what Herman Daly has referred to as the "invisible foot" that so often comes along behind. "Air and water are used freely by all, and the result is a competitive, profligate exploitation—what biologist Garrett Hardin calls the 'commons effect,' what welfare economists call 'external diseconomies,' and what I like to call the 'invisible foot.' Adam Smith's invisible hand leads private self-interest unwittingly to serve the common good. The invisible foot leads private self-interest to kick the common good to [pieces]."[15] In our worship of "efficiency," furthermore, we sacrifice not only our environment but "time intensive activities (friendships, care of the aged and children, meditation, and reflection) . . . in favor of commodity-intensive activities (consumption)"[16] and we narrow our attention to profit and material interests.[17]

As Daly and others have pointed out, our gross national product (GNP) *rises* with increases in environmental *damage* if that damage is repaired through human efforts. The objective of economic growth (growth in GNP) is therefore served in some instances by energy policies that decidedly do not serve our collective interests, as they involve substantial environmental costs. Maximizing GNP and GNP growth, as Daly sees it, is equivalent to maximizing the "throughput" of energy and other resources from their sources in nature back to natural sinks as wastes. Such a goal does not maximize human well being. Daly proposes not only an effort to stabilize human populations but an effort to move toward other elements of a "steady state" economy as an alternative to GNP growth. Our objective, he suggests, should be to maximize the service we get out of a given throughput. He breaks this objective down through the following formulation:

service/throughput = (service/stock) $\times$ (stock/throughput)

For a given "stock" of goods, equipment, and people, our objective should be to maximize the service we get out of that stock and the stock we are able to maintain from a given throughput. Although a focus on GNP often amounts to an effort to maximize throughput, a concern for the sustainability of our patterns of life should lead us to efforts to *minimize* the throughput associated with the services it provides. The quantitative details of an alternative measure to GNP are complex and, in many contexts, unnecessary, but specific proposals have been advanced.[18]

Daly argues further that we too often allow economic decisions about intermediate means and ends to eclipse broader thinking spanning the range from "ultimate means" to "ultimate ends." Where physics and the engineering sciences might more appropriately be applied to an open exploration of natural relationships and ultimate means, and where religion and ethics might come into play in the formulation of ultimate ends, we tend instead to focus on very narrowly defined choices—even "to take our conventional priorities as given and then deduce the nature of the Ultimate End as that which legitimates the conventional priorities."[19]

The supply-oriented energy policies of the recent past have been very much throughput oriented and may very well provide the best short-run boost in GNP statistics. Traditional discounting assumptions also lead to production of a low initial cost, relatively short-lived stock of goods and equipment, including a great deal of energy supply and conversion equipment. This stock is, in turn, characterized by relatively high maintenance and replacement costs (*future* costs that appear small when discounted), also perhaps making GNP statistics appear most healthy. Daly and others like him simply suggest that these policies and practices may not best serve the common good.

From their arguments, it appears that conservation and renewable energy systems, in particular, may be worthy of adoption even where higher initial costs and long lifetimes (with proportionately delayed and discounted returns) make them unjustifiable by traditional standards. If service is to be maximized in a sustainable steady state economy, energy policies radically different from those suited to GNP growth may be in order.

Other Disciplinary Perspectives and the Radical Impoverishment of Imagination

Experts really begin to differ, of course, if we delve into other disciplines as sources of expertise. In the early days following the 1973 energy crisis, for example, the perspective of certain political scientists led to questions about the very reliance on experts in the first place. "Studies are needed on the extent to which formal theory can indeed point toward optimum policy, and those areas where analysis is powerless to direct or guide choice, *either* because the issues do not require primarily the orchestration of means but rather the choice of ultimate social ends, *or* because ostensibly 'value-free' analysis actually insinuates so many hidden assumptions into the choice process as, willy-nilly, to make it political rather than technical."[20] Far from accepting a priori the assumptions of economists and engineers, it was noted that even such things as the definition and prevalence of corporations as social institutions are, in a certain sense, matters of political choice. The institutionalized motivations, accounting practices, and decision making procedures of corportions constitute, in their energy effects, energy "policies" that are from this perspective as much subject to political adjudication as any other energy policy.

Historical and anthropological perspectives also raise profound questions about the very frameworks employed by other experts in addressing energy issues. Historian M. I. Finley, for example, has described ancient Roman civilization as one in which modern economic concepts did not govern social or individual behavior and would in fact have been fundamentally incomprehensible. Referring to Alfred Marshall's book, *Principles of Economics*, published in

1890, Finley states that, "Marshall's title cannot be translated into Greek or Latin. Neither can the basic terms, such as labour, production, capital, investment, income, circulation, demand, entrepreneur, utility, at least not in the abstract form required for economic analysis. In stressing this I am suggesting not that the ancients were like Moliere's M. Jourdain, who spoke prose without knowing it, but that they in fact lacked the concept of an 'economy', and, *a fortiori*, that they lacked the conceptual elements which together constitute what we call 'the economy'."[21] In a society "not organized for the satisfaction of its material wants by 'an enormous conglomeration of interdependent markets,'" it was, not surprisingly, impossible "to discover or formulate laws ('statistical uniformities' if one prefers) of economic behavior, without which a concept of 'the economy' [was] unlikely to develop, economic analysis impossible."[22]

According to Finley, wealth was important in the Roman world primarily because of the "conviction that among the necessary conditions of freedom were personal independence and leisure."[23] Agriculture, as the fundamental source of wealth and freedom in this sense, was entirely revered over the crafts, trade, or other forms of commerce, even where the latter might in modern economic terms be equally profitable to the individuals involved. Ownership of land was, in keeping with these values, an exclusive prerogative of citizens. Wage labor was quite rare. Money lending occurred as a form of political alliance, often involving the purchase of land, but where interest was charged at all, rates varied widely. Division of labor was seen as a means to an improved quality in the product, not a means to quantitatively increased outputs. Major cities, far from being centers of commercial economic growth, were parasitic upon the agricultural base, and existed primarily as the political centers of consumption by free men with major land holdings.

In these and other respects, Finley's picture of ancient Roman civilization provides detailed evidence for his statement that, "Economic growth, technical progress, increasing efficiency are not 'natural' virtues; they have not always been possibilities or even desiderata, at least not for those who controlled the means by which to try to achieve them."[24] As depicted by Finley, the marked Roman preference for household self-sufficiency through agricultural enterprise not oriented toward growth in a modern sense, and the very different social and political institutions and relationships of the ancient economy, entailed desiderata so different from those of our own society as to be difficult to understand or imagine from a modern perspective.

The point here, of course, is not necessarily to endorse Finley's interpretation of Roman times or to exalt Roman civilization, with its institutionalized slavery and other arguable flaws, as a model for modern times. Neither does Finley make any attempt to apply his interpretations of Roman times to the energy issues of interest here. This and other historical examples, however, cast our own worldview in a different light and begin to show the degree to which our energy decisions rely on culture-based assumptions that are less than absolute, less than invariant over time. The measures and objectives, the very concepts, we employ in the analysis of energy alternatives today represent choices at the fundamental level defining our society and culture as it exists today. But they do *not* constitute defining features of human culture or society as it has always been or necessarily as it always will be and they are *not* necessarily constitutive of culture or society as we may wish to define them in our own future.

Equally startling contrasts can be drawn from the field of anthropology. In his book *The Original Affluent Society*, for example, Marshall Sahlins[25] describes societies in which the accumulation of goods inhibits essential mobility and in which "affluence" and the "good life" may be more nearly assured by underproduction than by unrestrained effort. Employing what Sahlins terms the "domestic mode of production," these societies seem content to produce for livelihood in intrinsically antisurplus systems. Production for use value eclipses production for exchange value and ritualized exchange systems often serve more to affirm social relations than as a form of competition or avenue toward accumulation or profit in any Western economic sense. People in the societies Sahlins describes are most especially "affluent" in terms of the sizable amounts of leisure time they have, an affluence that seems to belie all pretense that the frantic pace of industrialized economies is specifically geared toward freeing its workers' time from toil.

In his descriptions of hunter-gatherer and other societies, Sahlins notes that "there are two possible courses to affluence. Wants may be 'easily satisfied' either by producing much or desiring little." In describing an "original affluent society," Sahlins denies "that the human condition is an ordained tragedy, with man the prisoner at hard labor of a perpetual disparity between his unlimited wants and his insufficient means." Sahlins and others[26] suggest that "it was not until culture neared the height of its material achievements that it erected a shrine to the Unattainable: *Infinite Needs.*"[27]

Here again, the object is not to hold up any particular society as a model but to illustrate the radically different images of human behavior and, by implication, the sharply different ranges of human possibility envisioned by "experts" from different disciplinary backgrounds. Even a half-baked appreciation for the staggering richness and complexity of human experience, as explored through disciplines such as history and anthropology, would seem adequate to shame the simplistic behavioral assumptions of the engineering and economic responses into full retreat. Their deep appreciation for the range of possibility, however, may be a major factor, itself, in discouraging experts from these other disciplines from becoming involved in energy policy debates. Their nonparticipation may, in turn, help to explain the radical impoverishment of imagination evident in the usual engineering-economic assumptions adopted without challenge in works like the CONAES report.[28]

The remarkable inclusion of an anthropologist in one of the CONAES background study panels, in fact, produced the following observation: "anthropologists study the past and the present; we don't study societies that don't exist, nor do we invent them. I soon learned that our humility was probably misplaced in [working on the CONAES project, however] because economists don't mind inventing all kinds of societies."[29] Although less assertive in applying their knowledge base to practical decision making on energy issues than engineers and economists, experts from other disciplines may have a great deal to offer in grappling with energy futures, even in the absence of unabashed and unambiguous guidance with respect to immediate regulatory and technological choices. From many of these perspectives, modern economic desiderata, in particular, can be seen in large measure to be artifacts of a particular time and culture—abstractions, perhaps, from a practice that has by certain measures been extremely successful—*not* by any means absolute, complete, or permanent guides to a desirable future.

So Experts Differ

If we consult widely, it seems that doubts and questions begin to proliferate almost endlessly. Looming ever closer on the horizon is the specter of the "deconstructionist's" nightmare of undirected bits and pieces with nothing sure or secure.

Traditionally, of course, we escape such nightmares through a kind of "expert poll." Surely there are many more economists and engineers who accept, apparently unquestioningly, the guides they have been taught to employ than there are dissenters in these or other disciplines. With few exceptions, in fact, dissenting experts are relatively obscure, and if we do not actively ferret them out and cull through their arguments, we can well remain oblivious to the concerns they might wish to raise. Those from disciplines like history and anthropology almost never specifically address such issues as energy policy anyway. And the combination of their nonparticipation and the assertions of those employed in the "relevant" disciplines (engineering and economics) may persuade us that the expertise of historians and anthropologists in fact cannot be brought to bear in such matters.

Yet how reassuring can it be, for example, that the once active proponents of "net energy analysis," an alternative analytical procedure comparing alternative courses of action on the basis of their net energy, rather than simply economic, effects appear effectively to have been silenced by disapproving economists.[30] Does this outcome ensure that the economists were right or that those concerned with the potentially unique physical significance of energy as a input to production were wrong? Or has this outcome been more a function of the numbers of experts holding each view, the security of their funding or institutional positions, or the gradual post-energy crisis dissipation in public awareness of the importance of a conscientious resolution of the debate?

Can we expect "truths," to the extent that we can talk about such things, to be the easy products of a democratic process among experts? Can we even expect the process that occurs among experts to be "democratic"? Or does the polling process we implicitly accept restrict the sample of experts to be consulted in a manner that is suspect and give excessive weight to large numbers of experts who may, in fact, be *dis*advantaged by many years of narrowly shared disciplinary training? To what extent, in other words, are these debates being resolved on purely technical grounds, and to what extent might political or other factors play a role?[31]

How are we to know whether the dissenters or potential dissenters are "crackpots" or prophets ahead of their times? How can we tell whether the views of the dominant expert majority are solid and accurate or simply the modern day equivalent of the emperor's new clothes? For better or worse, neither the number nor the position of those pontificating can be fully reassuring.

FIVE

The Shaping of Responses

At this point we have raised a great many questions about traditional approaches to energy issues and noted the wide range of unknowns implicit in debates between experts from various fields and perspectives. Yet our discussion to date has been a bit abstract and theoretical. Is any of this of actual significance, one might ask, or is it all ultimately epiphenomenal: are energy decisions not actually made through what amounts to a complex real world political process? How are our responses to energy issues actually shaped in real life, and should we not be focusing our attention there?

If we adopt Harold Lasswell's classic definition of politics, that is, a process by which we determine "who gets what, when, and how,"[1] then energy-related decision making in practice is undoubtedly a matter of politics. Yet the theoretical abstractions that gain truth value in society generally play a significant role in politics.[2] In the politics of energy decision making, for example, popular acceptance of engineering-economic treatments gives these versions of truth great influence. In energy matters, it is difficult to separate practical politics, abstractions taken to be true, and actual palpable realities, all of which tend to be inextricably and significantly intertwined. Still, because the realities are always a bit beyond our grasp and the abstractions have now been attended to in some detail, it is perhaps time to go on and consider the practical politics involved in the shaping of energy responses.

With this chapter, we begin moving in the direction of outlining alternative solutions—or more accurately alternative approaches—to energy concerns, although the purpose here will still be more to raise than answer questions. By this point the reader may be getting impatient with the process of raising questions. What, after all, is the point in questioning assumptions and pointing out uncertainties if you cannot answer the questions you raise more convincingly than those whose work is being critically challenged? Because we must move ahead in *some* fashion, it may be felt that the real burden is to demonstrate that there is a better way to go than the present way.

Maybe so, maybe not. Perhaps the questions themselves are actually more useful than anyone's particular set of answers in spite of how uncomfortable they may make us. Well founded *dis*comfort is preferrable to *un*founded comfort, at least in the sense that discomfort may help to stimulate independent critical thought and active engagement with the problems at hand.

In any case, this chapter will offer an alternative set of assertions or interpretations as to what we might make of the politics of the situation.[3] The benefits of such a consideration of the practical shaping of responses to energy concerns are twofold. In the first place, an examination of the shaping process seems to confirm the arbitrariness of present practices and make it even more clear that our present response is by no means the only natural or logical response, by no means necessarily the most appropriate response, no matter how widely endorsed it may appear to be. In the second place, an examination of the shaping process can help us to understand both how we got where we are now and how we might begin to move in new directions. Understanding this process may be the first step in formulating new approaches (the closest thing to solutions this book will offer) to energy problems. These themes motivate the characterizations of traditional and alternative responses and of power and responsibility that make up this chapter. It is left to the reader to probe his or her experience and determine whether or not these interpretations make sense.

Collective Momentum

In practical terms, of course, our response to energy concerns is shaped by factors and processes far too complex to treat exhaustively here. We can state with certainty that this response is not simply the product of a direct popular democratic process; we do not, for example, suspend or reactivate construction of particular power plants or funding for fusion research on the basis of daily polls, (although to some utility executives and researchers it may seem that we do). Neither is our response a product of definitive technical calculations as the contents of previous chapters should make clear. Exactly what the process is, however, remains a murky issue with arguments that might be marshalled in support of a variety of interpretations.

It is my belief that the factors most significantly shaping present responses to energy concerns are actually neither democratic nor technical in nature. Present responses appear to stem instead from a widespread failure even to recognize and engage the issues directly. Specifically, our responses seem to be products of *institutional or collective momentum*, on the one hand, and *individual or popular nonparticipation*, on the other. This position is quite distinct from the technological determinist's view that science and technology taken collectively have acquired an inexorable logic of their own and now advance with humankind as irrevocable slave rather than master. It comes much closer to what Langdon Winner has so aptly termed "technological somnambulance," our sometimes mystifying tendency in the development and adoption of new technologies to "sleepwalk through the process of reconstituting the conditions of human existence."[4]

A brief but eloquent illustration of what is meant by institutional momentum can be found in a recent study at the Massachusetts Institute of Technology's Energy Laboratory,[5] in which researchers examined a variety of methods for removing carbon dioxide from the air emissions of electric power plants. Stimulated by concerns about the possible global warming effects of continued carbon dioxide emissions from fossil fuel combustion at power plants, the MIT study considered four approaches to removing CO_2 from stack gasses. Once removed, the CO_2 would be pressurized and piped into the deep oceans, down to a depth of 700 to 1000 m. (half a mile or so). Dispersed there, it would dissolve before reaching the surface, ultimately remaining in the deep ocean rather than contributing further to increases in atmospheric CO_2 concentrations.

The indications of institutional momentum at work show up in apparent inconsistencies in the laboratory's attitudes toward competing means of addressing possible CO_2 emissions problems. In a three-page summary of the CO_2 removal study,[6] the laboratory notes that: "Other options [for reducing CO_2 emissions] such as geothermal energy . . . and solar-generated electricity are environmentally attractive." In operation, they would generally release no carbon dioxide because they produce energy without burning fossil fuels. The summary dismisses these alternatives, however, on grounds that "their technology is not yet sufficiently developed for them to provide widespread, commercially competitive baseload power." Yet the same summary reports that the CO_2 removal technologies under consideration at the laboratory would roughly *double* the cost of electricity production. Carbon dioxide and water vapor are, after all, the predominant products of the combustion process in which the carbon and hydrogen making up the fuel combine with oxygen in the air. Removal and disposal of the vaste quantities of CO_2 produced at a power plant constitute an engineering task roughly comparable in scale to the power production process itself.

The laboratory summary concludes that, although the cost of its CO_2 removal systems "now sounds exorbitant, it may come to seem acceptable if the alternative is significant global warming." Yet, almost in the same breath, geothermal and solar alternatives for reducing CO_2 emissions have been dismissed as "not yet sufficiently developed . . . to provide . . . commercially competitive baseload power."

This apparent inconsistency may seem incomprehensible at first; these are, after all, intelligent people. But from an institutional perspective there may actually be no inconsistency. MIT is, after all, a research university. It is not a private corporation in the as yet little developed business of installing photovoltaic systems on people's roofs or building small geothermal power plants for municipal utilities. It employs and trains large numbers of highly specialized scientists and engineers who often work with other specialists for large organizations on large engineering projects. Its Energy Laboratory, furthermore, obtains much of its funding from electric utilities, major oil companies, and other large corporations with major investments in coal-fired power plants and other fossil fuel-based systems. Even with the most dedicated effort, this institutional assemblage could neither willingly nor by force adopt wholly new objectives overnight.

Carbon dioxide removal is a problem well suited to MIT's structure, funding practices, and capabilities as an institution. Adoption of existing geothermal and solar alternatives offers MIT and its usual associates little or no real role. (Continued research and development of "improved" renewable technologies, on the other hand, has been of interest at MIT.) By nature, the institutions represented here have developed a kind of collective momentum[7] and their handling of alternatives for CO_2 emission reductions is eminently consistent with that momentum.

An equally telling but more general example of the effects of momentum can be found in the early history of U.S. responses to the energy crisis, as this response was thrown into question by the work of a lone researcher, Amory Lovins.

Early energy studies, whether backed privately[8] or by the government,[9] almost uniformly adopted assumptions calling for growth in energy use at or near historical rates. These studies, and the government policy they supported, focused on increasing nuclear and fossil fuel energy production to meet growing energy "needs" and on long range research and development for breeders, fusion, unconventional oil and gas recovery, and other high-technology systems to provide for continuing growth in energy use into the indefinite future. These were, essentially, extrapolated expressions of existing institutional tendencies and the interests and capabilities of the "logical" energy experts in existing electric, oil, gas, coal, and nuclear industries and related academic disciplines.

The effects of momentum came into sharp relief with the publication in 1976 of Amory Lovins's classic article, "Soft Energy Paths," in

Foreign Affairs.[10] In this article, Lovins, a young, independent analyst (British representative for Friends of the Earth), pointed out that a whole realm of energy alternatives had yet to be broadly considered.[11] There were, after all, technological opportunities for increasing the efficiency with which energy was used. And a whole collection of renewable, largely solar, energy alternatives were not being tapped in the traditional supply plans. Lovins further argued that official plans for responding to the energy crisis were infeasible on technical grounds. Projected increases in electric power production, to be accomplished by rapid construction of coal and nuclear power plants, would absorb virtually all available investment capital, leaving little or nothing to build the facilities that were to use the electricity. Basic resources like cooling water would also be insufficient for such a program which could by the year 2000 be "responsible for the release of waste heat sufficient to warm the entire freshwater runoff of the contiguous forty-eight states by 34–49°F."[12] Federal agencies and private enterprise both, according to Lovins, had "wholly ignored both price elasticity of demand and energy conservation" in their plans and projections.[13] Lovins called for a *reduction* in electrification rather than an increase, and for a host of efforts to develop cogeneration and other efficiency improvements. He argued that technical efficiency improvements and renewable energy supplies could, over time, supplant conventional fuels in a cost effective manner and that these possibilities constituted a "soft path" alternative to traditional, growth-oriented "hard path" energy futures.

Lovins's proposals contributed to a fire storm of popular debate lasting over a period of years. Popular support for his soft path proposals was an important factor in the politics of the first "Sun Day" in 1978, in the first significant funding for solar research at the new Solar Energy Research Institute, and in the adoption in 1977 of federal tax credits for homeowners installing small solar energy systems.[14]

Although there remains room for debate about the precise requirements of a soft path future, Lovins's proposals helped to reveal a staggering gap in official analysis of energy options for the future. Analysis as well as actual policy had apparently been blind to a whole range of alternatives simply because they departed from the conventional wisdom of the time and the existing distribution of interests responsible for that conventional wisdom.

Other, more subtle expressions of similar forces shaping our collective response to energy concerns have been neatly captured by anthropology professor Laura Nader from her experience working on the CONAES[15] study mentioned in earlier chapters. Noting what seemed to be an absolute taboo on the word "solar" in the CONAES committee's proceedings, she reports,

> Their memos discussed nuclear, coal, and non-nuclear. Non-nuclear was solar.
>
> I asked the co-chairman, "How come nobody ever uses the word 'solar' around here? I've been on board six months and nobody's used the word 'solar'." He looked at me, rather surprised. "I don't know. Solar's been an orphan child." Somebody else piped up. "Solar? Solar's not very intellectually challenging." Somebody else said, "What's solar? A bunch of mirrors."[16]

For an anthropologist, these are fascinating remarks. Why, Nader asks, is solar an "orphan child"? Does this, for example, have anything to do with World War II and the development of nuclear weapons, the subsequent history of nuclear power development, or residual military influences in the shaping of energy policy? "The other observation: 'Solar's not very intellectually challenging.' What is intellectually challenging to these people? They seem to relish something complicated, hazardous, difficult and risky, something that requires high technology and big money."[17] The shaping of our energy futures by major social institutions might well occur, then, not only through singular decisions, but over extended periods of time. It is a process that, at a minimum, intersects with the socializing effects of our educational system and reflects both the way we organize knowledge and our dependence on a system of intellectually isolated disciplinary specialists.

More than institutions like MIT alone, the accustomed behavior of a whole array of mutually articulated social institutions tends to acquire a momentum of its own, carrying past practices and arrangements into the future. As citizens we do, undoubtedly, have some influence over the directions of basic research and over such things as the budget at the National Science Foundation. Our elected representatives at least have nominal control of direct government expenditures. But, in the absence of major social uprisings, those who serve on grant review committees, government policy advisory panels, and major corporate boards of directors generally are in a better position to stay the course in shaping our collective response to energy concerns than ordinary citizens are to bring about changes. The line engineer with his or her trained notion of what is and is not an "intellectually challenging" project, may also be better situated than is the ordinary citizen to shape our energy future. And, of course, public and private leaders, line engineers, and most citizens have had their thinking powerfully shaped by the social and institutional setting with which they have experience. From the top to the bottom of the hierarchy, the active players generally have every immediate reason and incentive to move in ways that reflect the structure and interests of the major institutions they represent. Where there is any doubt, the rationalization that "what is good for me, my family, and those I know must be good for the nation and society as a whole" has to be one of the most irresistable of all rationalizations.[18]

It is true that during the crisis periods of the 1970s and early 1980s, there were some signs of political realignment. Public attention was focused on energy problems to an unprecedented degree and congressional action produced such apparent aberrations as the tax credits for implementation of conservation and renewable energy systems. (These tax credits were abolished in 1985.[19]) Even during this period, however, programs more consistent with established distributions of interest effectively predominated, from funding for synfuels and nuclear power to the policy planners' minimization of the long-term contribution that conservation and renewables might make to the nation's energy systems. The election of Ronald Reagan in 1980 signaled an end even to the brief period of realignment, as his administration effectively dismantled the Solar Energy Research Institute (SERI), redirecting it with sharply reduced funding from short-term commercialization efforts toward long-term research. Since the early

1980s, and especially after the decline in real oil prices after 1986,[20] energy policies have drifted ever closer to what one might expect as an expression of collective momentum. At this writing, there are hints that continuing distinctions may be possible between Democratic and Republican administrations with respect to national energy policy. But departures from the momentum of the past appear even less likely under Clinton than they were under Carter, unless crisis somehow again brings public attention to a focus.

Present energy policies[21] emphasizing continued use of fossil fuels along with sustained efforts to develop satisfactory nuclear power options are certainly more consistent with institutional arrangements dating from well before the 1973 energy crisis than any alternative tilt toward conservation and renewables, for example, could be. Both present policies and our continued reliance on market economic language to justify those policies, can be seen simply as *reflections* of those institutional arrangements. They are neither explicitly democratic nor fundamentally technical. They simply reflect preexisting distributions of established interests.

It should perhaps be emphasized that, for the time being, the intent is *not* to argue that CO_2 removal will never prove useful. Nor is it argued here that other pet projects well suited to the needs of an organization like MIT—projects such as their work on a new generation of "inherently safe" reactors—are necessarily ill advised. Neither is it clear that existing geothermal or solar alternatives will prove less than twice as expensive as conventional power production in all instances or that this is an adequate criterion for judgment. The point here is simply that the institutional structure and organization of society as reflected in the decisions and actions of individuals and groups have been among the most powerful (though often among the most subtle) forces shaping what research has and has not been done, what technologies have and have not been developed, and what energy systems have and have not ultimately been put into place.

To this last statement, one may respond in disgust, How could it be otherwise? Indeed. The importance of the observation, again, lies in its implication that the particular biases introduced in our decision making by this collective momentum and the specific mechanisms by which these biases are introduced may be worth monitoring. In keeping track of where choices are possible and where we must simply yield to inevitable realities, we need to remember that *institutions can be changed* and that *societies can be variously organized.* In monitoring the specific effects of present structures we may wish to consider how else our responses might be patterned and where the differences might lead us.

Popular Nonparticipation

Our little example of MIT's rather odd handling of alternatives for reducing CO_2 emissions is interesting in another respect: it may be indicative of the extent to which we as a society defer to technical experts in such matters and, in this sense, of another important element in the shaping of responses to energy concerns. Perhaps this takes the MIT example too far. But in combination with the apparent circularity of some economic reasoning and other instances of less than robust logic mentioned in earlier chapters, it begins to look as if the experts involved may not actually *expect* to be called to account, even where they engage in logical fallacy. Within the circle of expertise, there is a shared notion of which alternatives are and are not "challenging," of which assumptions are and are not "reasonable," and the matter is presented as if popular deference to expert judgment were simply to be assumed. The role of institutions like MIT in shaping our energy future must be considered not simply by examining the institutions themselves, but by reflecting on popular response as well. The influence of an institution like MIT may be much enhanced not only by its links to other public and private institutions but also, in this case, by an institutionalized deference in our society to precisely the kinds of expertise MIT represents.

This brings us to the second of the two most important forces shaping our response to energy concerns: individual or popular nonparticipation. Nonparticipation is evident in virtually all consumer behavior, in the popular failure to ask and demand answers to the kinds of questions raised in this book, and in our generalized acquiescence in accustomed patterns of behavior. We do not ask why the refrigerator we purchase uses 3–5 kWh a day rather than one tenth that much. We do not ask why builders continue to build conventional homes when well-demonstrated "superinsulated" designs no longer require heating. We do not press the question of what future generations will do for energy or how we expect to deal in the near or long term with record breaking dependence on foreign oil imports. Numerous studies indicate that most of us do not even behave "rationally" in the simplest case of an appliance purchase, walking out happily with a machine that will often cost us more for energy in its first year or two of use than we saved by buying the less energy efficient model. Most of us, one begins to suspect, simply are not paying much attention.

In particular, *we do not insist that policy makers and technical people seriously explore with us more than one set of sociotechnical alternatives.* Instead, we seem to accept, with little or no comment, the same old patterns dressed up with the most modest adjustments at best.

Ask people you meet on the street to explain to you how they would expect the adoption of nuclear power or of renewable energy systems to alter their or their children's pattern of life in the future. Ask them to describe for you an alternative to present patterns that does *not* include wall switches supplied by remote and anonymous electric utilities or does *not* include travel to and from work and play by automobile, with a well-supplied fuel station on every corner. Ask them even to give you a good estimate of the relative scale of their energy use for various applications—home heating, auto use, electric lighting—and how this use compares with their own physiological capabilities. What, in other words, is their current level of dependence on nonfood energy inputs and how might this facilitate or constrain their behavior in the future?

The widespread popular failure to probe these and other energy related issues constitutes a major ingredient of "technological somnambulance" with respect to energy choices. In practical terms, it is a major contributor to existing energy practices and policies. Those who do not think about such issues can hardly play an important *active* role in the choice process.

I do not mean to blame the victim, here. And it is possible to address energy issues indirectly, with little or no specific concern for kilowatt hours or physical quantities of any sort. The kind of analysis of decision processes underlying this chapter alone, for example, could lead one to active involvement in energy decision making without any direct familiarity with physical, economic, or other abstract notions.

Nonparticipation is understandable. However misplaced, many people do have a genuine sense of incapacity to deal with energy matters, a sense perhaps reinforced by the esoteric dialects and professional airs of the traditional experts, many of whom are themselves misled by the lack of basic challenges within their areas of expertise. Most of us are also far too busy trying to hold our lives together—multiple career families, the house, the kids—to wax philosophical about "energy futures." Although we live in one of the richest countries in the world, many of us are, or at least have been, convinced that what we want in this world is scarce;[22] constant stress and striving are the price we must pay and we simply do not have time to stop and look at the overall picture, whether centered in energy concerns or otherwise. Finally, and here we may not be let off the hook as easily, we may not be sure that we *want* to look. If we begin reexamining things at the roots, how are we to be guided? Will we perhaps discover that our own convictions require that we give up some of the comforts or amusements we now enjoy? Will the dream of success and security for our children be shattered by harsh new realities? There are a great many unknowns out there and in many ways it is more comfortable not to stick our necks out and begin asking questions.

What, furthermore, can one person or family do? We can vote in the November elections, of course, but the link to energy futures is remote and tenuous. Supposing instead that we decide to take action in our own lives, that we take the trouble, for example, to get a solar water heating system installed. Supposing we do take the risk, locate what we hope will be a competent installer, fill a small room with complicated networks of tanks and piping, work through the debugging and start-up problems, and yes, ultimately, pay the bill. What real good will that relatively straightforward departure from custom do in the end? Doing things differently is risky and difficult and likely to lead to all kinds of conflicts with established patterns. Solar water heating is one thing, full-fledged reliance on renewable energy systems would be quite another. And really stepping out and doing something that, for example, is not thought to be "cost effective," would require truly concerted effort, to say nothing of large amounts of time and an arsenal of convincing explanations for family and neighbors. How can we as individuals expect to come up with more sensible energy practices on our own in any case? And again, in the face of seemingly monolithic custom, what really would be the point?

No, nonparticipation is not particularly surprising. It does, however, appear to be a profoundly effective factor in the shaping of our collective response to energy problems.

Alternative Responses

There will be those who will argue—and they cannot be defeated absolutely—that what has been described as collective momentum and popular nonparticipation simply amounts to an expression of preference: people *prefer* present patterns and have in fact endorsed established energy policies. For me, this is a bit like arguing before the era of Mikhail Gorbachev and Peristroika that the citizens of the USSR endorsed the autocratic rule of the Communist Party and the repressions of the KGB. The lack of conscientiously defined alternatives in both cases makes it difficult to place much faith in such an argument.

It is clear, on the other hand, that unusual times and even modest openings in prevailing institutional arrangements have been accompanied by a collection of very unconventional responses.[23]

Popular support for the development of solar energy, for example, did become quite vigorous and widespread under the abnormal circumstances of the energy crisis. A number of opinion surveys conducted in the 1970s indicated that a substantial minority (47 percent in one case) of those polled would be willing to consider solar energy systems even at prices somewhat higher than those for conventional energy sources.[24] And, unable to get a response from established bureaucracies commensurate with popular support in the same time period, Congress was moved to form a completely new national laboratory (the Solar Energy Research Institute) to pursue solar research and development. Radically different positions also crept temporarily into official documents; widely quoted at the time, a report from the President's Council on Environmental Quality stated that, "A strong case can be made that a national commitment in the 1950's to develop solar technology—comparable to the one made to develop nuclear power—would have led to the widespread economic feasibility of solar energy today."[25]

Antinuclear sentiment also blossomed in the 1970s and 1980s with the Three Mile Island (1979) and Chernobyl (1986) accidents. Whether these incidents were in fact statistical aberrations or evidence of erroneous assessments of actual accident probabilities, they contributed to a popular rebellion against expert nuclear advice that throughout the 1950s and even into the 1960s had been accepted on faith. New plant orders dried up and a number of previously ordered plants, some even in advanced stages of construction, were cancelled. Scientific opinion also shifted as new, much higher estimates of accident probabilities were developed.[26] Both the Clinch River breeder reactor program and expansion of commercial nuclear power were effectively put on hold.

Perhaps most surprising, a seemingly innocuous new law, the Public Utility Regulatory Policies Act (PURPA), elicited a wealth of innovative small scale electric power production efforts after its passage in 1978 and gradual implementation at the state level over succeeding years. The new law simply required that major electric utilities accept power from small (nonutility) generators and pay a price for that power equivalent to the costs the utility avoided by not having to provide that power. State regulators were to ensure fair "avoided cost" rates for power from small producers who qualified under the new law, just as they oversaw rates for sales to utility customers. Over a period of only a few years, this new law created the opening for a revival of small hydroelectric power production, more than 1,000 MW of wind power in California, and rapid growth in industrial and other cogeneration (i.e., the much more energy efficient joint production of electric power and thermal energy in situations where electricity had formerly been purchased and thermal energy had been produced separately on site). Electricity began to be produced from small wood-fired power plants, from waste methane at sewage treatment plants and old landfills, from geothermal sources, and from pressure reduction stations in city water supplies. California, with a statewide generation capacity of only about 40,000 MW already had nearly 3,000 MW of nonutility power production in operation by 1985 with another 12,500 MW in various stages of development under signed contracts.[27] The pace at which these often highly innovative projects were successfully implemented was startling, especially in comparison to experience with large coal and nuclear plants (typically 1,000 MW each) in the same time period. So much alternative power production was coming on line in California that state regulators were forced to begin restricting project development by reducing avoided cost rates and instituting quotas by resource type.

Although collective momentum and popular nonparticipation may be the hallmarks of our response to energy problems, then, there is also ample evidence of departures from this norm. This evidence may be sufficient to conclude that our predominant tendencies are not the result of an actual lack of imagination or of an all-pervasive lack of willingness to take risks with innovative approaches. Neither is it consistent with the position that the lack of continuing vigorous challenges to present policies must be indicative of a correspondingly high level of support for those policies or low level of interest in alternatives. I know of no definitive way to determine when traditional responses may be appropriate elaborations of a successful and still desirable pattern of life and when they may represent a counterproductive buttressing of patterns already overdue for replacement. It is clear, however, that what may initially seem modest adjustments in the institutional framework, can elicit unexpected and markedly different responses.[28]

The "Promise of Technology"

How is it, if the preceding interpretation is accurate, that we allow our response to energy concerns with all their profound ramifications to be shaped largely by the politics of default—that is, by collective momentum and popular nonparticipation? As has been suggested in earlier chapters, the decisions that make up that response profoundly affect our relations with the natural world and with each other. In large measure, they define our pattern of life. How many of us would allow an economist to select our spouse on the basis of a cost benefit analysis? How many of us would allow an engineer to select a house for us, or even a car, by the same or any other method? Are these all choices with a more substantial personal component? Do they all, in fact, have more far-reaching implications for our relations with nature or with other people?

We live with pride under a Constitution explicitly designed to "establish Justice, insure domestic Tranquility, provide for the common defence, promote the general Welfare, and secure the Blessings of Liberty to ourselves and our Posterity"; yet in a wide range of public decision making we seem to accept arguments that fail to go beyond the discourse of corporate profit maximization. How is it that, in seemingly docile passivity, we so often remain on the sidelines as energy-related decisions are, in effect, made for us? Have we been hoodwinked by a powerful ruling elite? Are we simply unwilling to engage individually or collectively in a basic reexamination of the way we live our lives?

There is reason to suspect that the answer to both of the last two questions is a qualified "yes": there are elements of both *power* and *aquiescence to power* in the explanation. One might guess also that our present energy situation closely reflects the most fundamental beliefs and assumptions of a technological society, a hegemonic worldview,[29] in fact, that through gradual evolution has governed much of our thought and action for hundreds of years.

Would it even be reasonable to expect the traditions supporting choices made for us in the marriage of nature and mankind (a marriage implied in our energy systems) to be fundamentally different from the traditions that have long supported arranged marriages between people in other cultures? In both cases the mechanisms and implications of a hegemonic view are largely (and unsurprisingly) invisible. They are accepted as a matter of course in a setting that offers no clear conception of alternatives.

Yet with some effort, explanations may be derivable. In particular, an examination of collective assumptions that might loosely be termed our belief in the "technocratic concept of progress," or our belief in the "promise of technology," may offer valuable insights. These assumptions have been examined from a number of angles by a number of authors but let us focus on Albert Borgmann's notion of the "promise of technology."[30] Our default on energy issues, Borgmann might argue, is a product of our longstanding desire to believe in "the promise of technology" and our associated willingness to buy into the "device paradigm" of technological advance.

Based on the transformative power of the natural sciences and dating back to the time of the Enlightenment, the promise of technology (as Borgmann describes it) is a promise of "liberation, enrichment, and of conquering the scourges of humanity."[31] Originally focused on a liberation from starvation and disease, this promise has, in its modern formulations, essentially become the promise of a "technological fix." As that promise has been formulated in recent advertising, the difficult job of learning a language can be solved "thanks to a new electronic miracle," and it is possible to have some of the world's best culinary dishes "Without leaving home. Without waiting. Without cooking."[32] Indeed, an unspecified "life of greater fulfillment" is possible through "continuing advances in science and technology," according to the formulation of the promise provided by Jerome B. Wiesner in 1976 while he was president of MIT.[33] Steeped in the promise of technology, to say nothing of the fruits of its past achievements, how can we be concerned about an energy problem? Having bought into that promise, our *duty* must surely be simply to stand aside and allow technological development to take its benificent course. Surely it is not up to us to consider more expensive energy systems or less convenient patterns of life. We need only have faith and patience, and the development of new technologies will again bring us both cheaper *and* more convenient alternatives.

If we are believers in the promise of technology we may tend further to adopt a "device paradigm" in our thinking whereby we expect the benefits of technology to be uncomplicated. "[T]he notions of liberation and enrichment [that are part of the promise of technology] are joined in that of availability. Goods that are available to us enrich our lives and, if they are technologically available, they do so without imposing burdens on us." We may expect technology to deliver "devices" that supply the desired commodities or outputs while progressively removing the necessity of work, skill, or engagement of any kind with the functioning of the device. "In the progress of technology, the machinery of a device [is expected] to become concealed or to shrink. Of all the physical properties of a device, those alone are crucial and prominent which constitute the commodity that the device procures."[34] Operating under such a paradigm, it is a shock when electric utilities do not simply deliver electricity at the flick of a switch, but instead turn out to be occasionally unreliable, add unhealthy pollutants to the atmosphere, or threaten us with radiation or electrocution. Continuing within the device paradigm, however, we may yet await the next generation of technology which will surely deliver pure solutions, devices that do not burden us with complications but simply deliver the goods. Far from engaging the complexities of the energy issue, we expect and demand that the advance of technology make such engagement progressively *less* necessary.

Borgmann's notion of the promise of technology and its companion, his conception of the technological device, provide one way of describing the core of a hegemonic worldview of which we are at once victims and perpetrators. This fundamental, hence subtle and little recognized patterning in our own thinking in large measure preordains our response to energy concerns, unless we recognize and choose to rise against it. At the same time, it produces disproportionate benefits for those who play upon its unspoken assumptions, knowingly or otherwise. And it disproportionately penalizes those who chafe within its confines and do not share its values or delusions, whether or not they can express the source of their anguish and alienation.

Power and Responsibility

Quite apart from trying to discover how it is that we behave as we do, it is important to focus on the exercise of power implicit in our approach and consider the question of responsibility for the choices being made and their consequences. Without some consideration of responsibility and accountability, crucial motivations for more conscientious decision making may be lost. To pursue these issues, some careful attention to definitions will be necessary along the way.

Under classical definitions of power (in the political sense), A exercises power over B if A gets B to do something B would not otherwise have done. This definition does not, however, clearly cover a situation in which B's actions are altered by A's behavior without A's intent or knowledge. Under this definition, if public or private leaders bring about the adoption of energy systems that are not consistent with the values of the population and do not embody the kinds of relationships with the environment or among people that citizens would otherwise have chosen, it is not clear whether or not those leaders have exercised power over others if those others raise no active objection and propose no alternative plans. Similarly, if through nonparticipation, the citizens of today acquiesce in the establishment of energy systems and patterns of life that eventually lead to critical resource shortages or disasterous environmental conditions for future generations, it remains unclear whether or not those citizens have exercised power over future generations, if the actions of the former were taken in ignorance of the consequences.

Responsible decision making can be encouraged only under a more inclusive definition of power. In his recent book, *Power: A Radical View*, Steven Lukes[35] offers some essential clarification: "Can A properly be said to exercise power over B where knowledge of the effects of A upon B is just not available to A? If A's ignorance of those effects is due to his (remediable) failure to find out, the answer appears to be yes. Where, however, he could not have found out—because, say, certain factual or technical knowledge was simply not *available*—then talk of an exercise of power appears to lose all its point."[36] If the leaders or the citizens just mentioned *could reasonably have been expected to find out* about the effects of their decisions or actions on others, in other words, then their actions must be taken to be an exercise of power over others, whether or not they do, in fact, "find out." As exercisers of power, their level of responsibility and their accountability for the results of their actions are substantially heightened. If, on the other hand, we allow ignorance alone to guarantee their innocence with respect to the consequences of their actions, we lower substantially the standard of conscientiousness we require them to meet.

Our political thinking is increasingly corrupted by pluralist political and market economic models. The shared guide of caveat emptor would seem to excuse both the prominently placed effectors of collective momentum and the passive practitioners of nonparticipation from responsibility for the consequences of their actions. Leadership is excused: "If they objected, why didn't they say so?" And the citizen is excused: "How was I to know?" Each can readily point to the other. Yet together they are making the decision, even if only by default. Each is, in his or her own way, exercising power whether or not either one conscientiously explores the ramifications of that exercise. And each must, in this sense, bear responsibility for the consequences.

The experts, whether corporate executives or government officials, are not doing their jobs. By not questioning the status quo, by failing to explore conscientiously the possibility of alternative preferences, they continue to profit disproportionately and illegitimately from their positions and from past investments in a potentially outdated and inappropriate pattern of life. Ordinary citizens are not doing their jobs either. By failing to examine conscientiously their own actions and the implications of those actions for their fellow citizens and for the future, they too stand to profit disproportionately and illegitimately by avoiding the cost and complications of potentially needed changes. If this general diagnosis is on the mark we can look forward to a time in which there will be no shortage of people to blame. This opportunity may be small comfort, however, if it is obtained at the expense of highly undesirable energy-related outcomes that could more responsibly have been avoided.

SIX

Exploring the Option Space

What then should we do? What energy sources should we tap and at what rate? What technologies should we employ in energy conversion and end use? Over what time period, if at all, should we strive to make the transition to indefinitely renewable energy soures? What social, political, organizational, and institutional arrangements should we adopt in the production and use of energy and what arrangements should we avoid? What long-term relationship with the natural environment do we wish to express through all these decisions, and how should that relationship be implemented in practice?

If true democracy necessitates the close and active participation of citizens in addressing at least the questions that are of fundamental importance in defining the society of which they are a part, then these questions should not be answered by any restricted group or individual. Precisely because our present answers to energy questions appear to be somewhat arbitrary, precisely because vast realms of uncertainty and indeterminacy remain both technically and socioculturally, it is essential, in a democracy, that these questions gain currency and that broad participation be assured in arriving at answers.

What can be offered here, then, is not answers to the questions raised in this book, but a collection of tentative implications or alternative approaches to energy issues that may be drawn from the perspectives that have been presented. *The central theme in this regard will be a call for much more forceful engagement with the range of technical and sociocultural possibility inherent in energy decisions.* If our way of life in the future is to be democratically and actively chosen, the artificial barriers of conventional theory and practice will need to be recognized. But beyond this, if the full range of possibility is to be more directly explored, examined, and engaged, actual *experimentation* will also be required to gain experience collectively with a broader range of alternatives.

Most of this chapter will be devoted to an introduction to the "home power" movement, a small but expanding value-based social movement[1] incorporating extreme conservation measures and renewable electricity production in the homes of participants. This movement has emerged from very much the kind of experimental engagement with possibilities that is to be recommended here. In addition, it is one of the most advanced and coherent examples in our own society of a thoroughly distinctive set of answers to energy questions. As such it provides a concrete and graphic illustration of the extent to which both present and alternative answers to energy questions can define a way of life. While it is emphatically *not* offered as The Solution to our energy problems, the home power movement also poses specific challenges to traditional assumptions both about human values and behavior and about the range of technical possibility, thus illustrating in concrete terms many of the themes of this book.

Before turning to the home power movement, however, a brief preamble may be useful, covering the traditional expectation that more complete "solutions" be proposed in any treatment of energy problems worth its salt.

Appropriate Expectations

The questions raised in this book constitute an implicit if not explicit challenge to authoritative responses to energy problems and an encouragement to broadened participation in energy decision making. Whether fearing the chaos of broadened participation (not necessarily an unfounded fear) or wishing to defend the prerogatives of power, energy authorities often respond to this challenge with the charge that no superior alternative solutions have been offered. The questions that have been raised, they report, are not new to them (a claim that is at least partially true), but it does no good to throw out the old methods and solutions unless better alternatives can be proposed. To what extent are these counter charges legitimate? What is it appropriate to expect in the way of proposed alternatives before challenges to existing theory and practice can be taken seriously?

The authoritative response with respect to methodological challenges is perhaps most easily addressed. First, in questioning economic and other methods, the intent is to challenge not only the validity of the methods themselves but also the legitimacy of their use by economists and other experts. Calling not simply for an improved alternative economics, for example, the argument includes the position that democratic processes should displace expert methods. To require that challenges to traditional economic treatments of energy issues be accompanied by superior substitutes in the form of expert methods is to miss a major part of the point. Such expectations may be appropriate where expert decision making is to be preserved. But they are inappropriate where a central element of the challenge is the argument that the decision process itself needs to become more democratic. Similarly, if decision making is to be democratic, it is not enough for *experts* to be familiar with the problems plaguing traditional policy perspectives. *Citizens* also need to be aware that those problems exist.

But what is it reasonable to expect in the way of alternative solution proposals? Can we legitimately challenge present energy policies and practices if we cannot provide the details of preferable alternatives?

The answer here is yes. If we think carefully, the common call for alternative "solutions" in the sense of complete, full-blown alternatives to present patterns of life really is not reasonable itself. Energy production and use, after all, pervade every nook and cranny of our existence, from settlement patterns and food production and delivery, through industrial production, to entertainment at home (e.g., reading vs. TV) and on vacation (e.g., wind surfing vs. water skiing). When the very way we live is under scrutiny, it is too much to expect a complete portrayal of a full-blown alternative. Existing patterns of life, themselves, are far too complex and involve too many intricately articulated contingencies and interdependencies even to be described. If existing patterns themselves defy description, we cannot expect to perfect alternatives in the abstract, working out all their internal inconsistencies and resolving all the difficult conflicts of transition in advance. We can never realistically expect to offer fully developed options for the future.

If we persist in requiring complete descriptions of feasible patterns of life, the option that is closest to present realities will also tend systematically to win out over those involving proposals for change. Up to the point of its actual disintegration, after all, the present pattern is demonstrably "feasible," while any untried proposal involving changes from the existing pattern leaves room for doubt. Especially where we focus on short run thinking, the more restricted the departures from existing patterns are kept, the more "reasonable" the plan is likely to appear, right up to the point at which its systemic failure becomes incontrovertible.

As politicians, policy makers, researchers, and others concentrate on "reasonable" alternatives, they may also lose track of the fact that there are no alternatives for the future that do not in actuality involve untried change on a large scale. Population growth, conventional resource depletion, expanding degradation of the environment, and the inevitable implications of carrying on "business as usual" in the expansion of nuclear or other energy systems will bring fundamental and far-reaching change whether we like it or not. We do not have the option of preserving our way of life as it now stands.[2] Although the first step in traditional energy proposals may seem less foreign than the first step in new directions, the actual details of the future we commit ourselves to may actually be just as poorly understood in the former as in the latter case. We may feel that we know what life with nuclear reactors is like, but we may *not* know what life with a great many *more* nuclear reactors, including the essential addition of breeders and fuel reprocessing, would be like. Concerns alone about the civil liberties implications of dealing with potential or actual diversions of plutonium from fuel recycling—like the earlier mentioned routine surveillance of dissident organizations to ensure against plutonium diversion, or draconian and otherwise unlawful measures of search and seizure in the event of diversion—suggest the kinds of change that might follow what would otherwise seem a familiar first step.[3] Similarly, success with an electric automobile, although it might at first seem that it would preserve present patterns of life, would require nearly a doubling of present electric power production simply to provide the equivalent of present automobile energy use. Such a massive building effort, along with the parallel decline of present gasoline supply systems, would necessitate far-reaching changes in our lives even before we consider what fuel might be used in producing that additional electricity or how we intend to deal with continuing growth in auto use and congestion.

It is unreasonable to expect alternative "solutions" fully formed and delivered on a silver platter. It is incorrect to regard traditional policies as offering any guarantee of "solutions."[4] *All* of our options at present are little more than elements of partially formed directions for the future. And the shaping of our future that is implicit in present energy policies begins with first steps whether these be the design of "inherently safe" reactors or the installation of sulfur emission controls on coal-fired power plants. Those first steps will lead on to other problems, constraints, and decisions, none of which is yet necessarily more fully defined than they are for many alternative energy proposals.

We must recognize that either requiring fully developed alternatives before opening debate or falsely construing "no change" to be a secure fallback option is simply to endorse what has been described as our tendency to default in answer to the fundamental questions raised by energy choices. A refusal to give serious consideration to less than fully formed alternatives for the future *may* be an expression of preference for the *changes* implied by conventional energy policies. It may, on the other hand, represent a failure to recognize or acknowledge the changes that are implicit in traditional policies. Or it may reflect an implicit instruction from established interests that their objectives are well served as things stand and that the hoi polloi are simply to stay out of these issues, thank you very much.

In thinking about how we might improve our approach to energy issues, then, it is important to be careful about what we expect and why. And it is important to be aware of how our specific responses may be shaped, even in their earliest stages.

A brief diversion to the recent history of recycling efforts should help clarify what is meant here. For some years many people have felt that the recycling of paper, aluminum, and other materials would make good sense, even though it has not generally been economically attractive. Beginning to proceed on faith without a complete plan, we have found that all kinds of complications have arisen. Efforts to separate recyclables at traditional mixed waste disposal sites first presented challenges that remain to this day unresolved. From there we have moved instead toward separation at the source, but this has meant separate bins for glass, aluminum, recyclable paper, trash, and so on. These take up substantially more space than the simple single waste cans. And collection vehicles must make separate collection runs or include separate compartments on a single truck. Recycling simply does not fit in conveniently with the way we are used to doing things, nor is it as easy as the old way. Examined as an option for the future, recycling did not seem to make sense from economic and other perspectives; even well into the effort on faith, new problems seemed to mount.

And yet, over a significant period of time, the bugs may be beginning to be worked out. Oddly enough, workable solutions are also coming at a time when traditional disposal is becoming increasingly constrained as landfills are filling to capacity and convenient new landfill sites are found to be unavailable. Savings on traditional disposal costs associated with recycling have rapidly become a major benefit of recycling. In addition, the more recycling that has gotten under way, the better developed the infrastructure for recycling paper, plastics, and other materials, itself, has become; economies of scale for collection and transport have begun to emerge along with innovative applications for recycled materials and expanding markets for products. These factors together begin to make recycling more cost effective by traditional standards. But more important, for our purposes, what began as a seemingly irrational intrusion into relatively smoothly operating systems is gradually beginning to take shape as a part of new, increasingly smoothly operating substitutes. New complications, such as a mismatch in the pace of development of recycling programs and facilities for utilizing the collected materials, will continue to arise. But the initially "irrational" recycling effort may progressively become part of an alternative practice at least as "rational" as the practices it replaced.

"Solutions," in other words, may not be fully describable in advance, yet prove acceptable or even preferrable in the long run. There may be too many complicated interrelationships to permit a clear view of where we are going in advance, even with relatively limited recycling efforts, to say nothing of large scale alternatives to present energy practices. Yet it may be appropriate, and in the long run advantageous, to begin trying out some changes. If we are to make changes, and inevitably we are, then we must begin somewhere, even though the full implications of the changes we select remain unknown or unknowable.

The following discussion of the home power movement, then, is offered not as a panacea or even necessarily as a partial solution to our energy problems, but primarily as an illustration of the vastness of the "option space"—the technical and sociocultural terrain available for exploration and experimentation—and of the kind of experimental approach that could be used to explore and develop a wider range of alternatives for the future.

The Home Power Movement: Introduction and Technology[5]

The home power movement in the United States consists of a small but growing number of homeowners who have installed their own home electric power generation systems. These systems typically rely on lead acid battery storage and photovoltaic (PV) panels, often augmented by micro hydroelectric or small wind-power production. The great majority of the systems are not connected to the electric utility grid and are installed in homes that are located some distance from utility lines. Rapid expansion of the home power movement has occurred over the last several years, in spite of the expiration of solar tax credits, bringing total participation in the United States to the level of approximately 25,000 homes as of late 1989 and perhaps

100,000 homes by 1993.[6] Perhaps because of the kinds of people involved in the movement and because it has emerged from isolated individual efforts rather than from any centrally organized, monitored, or coordinated plan, the movement remained little recognized and largely unacknowledged in policy circles throughout its early years.[7]

Traditional approaches to energy issues assume that existing patterns of life are close to optimal and that marginal adjustments, such as switching to a slightly cheaper fuel or adding a bit more insulation to the attic, are essentially all that analysts need to consider. Yet home power adopters appear to prefer and have purchased energy supply systems and radical efficiency improvements beyond any simple rational response to energy prices. They appear to prefer patterns of life involving a greater variety of work activities, a strengthened sense of community, and distinctive relationships with the natural environment, which set them apart from traditional patterns in ways not predicted and not fully comprehensible under the usual expectation that change will occur only at the margins within an essentially fixed pattern of life.

To provide an understanding of the movement and its possible significance, this and the following two sections will describe the configuration of the renewable energy systems that have been put into place, the movement's historical development, and the apparent motivations of participants, respectively.

With respect to technology, home power systems cover a wide range of sizes and configurations. At one extreme are the systems that include minimal electric power supplies from one or two PV panels (say 50 to 100 peak W) with battery storage. Total electricity production from such systems would be on the order of several hundred watt hours per day, between one tenth and one hundredth the amount of electricity typically used in a home. Small systems of this sort generally provide energy for lighting alone or for lighting and communications (TV and stereo) and cost a few hundred to around a thousand dollars. Though perhaps most widely encountered on Indian reservations, very simple systems of this sort are not uncommon elsewhere, having been observed, for instance, in an underground home in the woods of central Minnesota, in lake shore homes in Michigan, and in community groups such as one near Tonasket, Washington. In the Minnesota case, minimal wood heating and the inhabitants' staple diet of dry goods (obviating the need for refrigeration) essentially eliminated energy purchases, but a reliance on bottled gas for cooking and refrigeration more often augments such small PV systems.

At the opposite extreme are systems with very large PV arrays supplying traditional collections of appliances. Although such expensive systems are welcomed by retailers and installers as an opportunity to sell large numbers of PV panels, they represent only about 1 percent of the total number of home power systems.[8] The vast majority of the systems include between 2 and 20 PV panels (at 30 to 50 peak W per panel), depending in large part on whether or not water-supply pumping and refrigeration are included among electricity uses in the home. With some obvious variation according to size and geographic location, these systems still provide only on the order of one tenth the electricity used in a conventional home.

PV panels appear to be by far the preferred and most widespread source of electricity for home power systems, but small wind generators (a typical rating might be 75 W in a 15 mph wind) and micro hydro systems (automobile alternator scale) are also widely employed to augment PV output and take advantage of seasonal variations in sun, wind, and precipitation. Relatively still, sunny summers and relatively windy but overcast winters make PV and wind a good combination at a site in northern Idaho, for example, while the combination of PV and micro hydro appears most attractive at another location near Cave Junction, Oregon, where sunny summers alternate with wet winters. There continue to be a few home systems (such as D. Harris's micro hydro home near Davenport, California, or E. Negro's wind-powered home near Mohawk, Michigan) that do not include PV power supplies, but these are now rarely encountered. PV panels, which conveniently have no moving parts, are generally included even in Michigan's Upper Peninsula where insolation rates are about as low as in any region in the contiguous states.

As one might guess from the small scale output of most of the systems, participants in the home power movement have by no means restricted their attention to the supply side of energy issues. They have, in fact, become undoubtedly the most sophisticated of residential energy users when it comes to efficient use of electricity. PV panel costs alone, exclusive of battery storage, charge controller, inverter (if 110 volt ac appliances are to be used) and other costs, amount to $5 to $10 per peak W.[9] On a life-cycle cost basis, it is generally said within the movement that adopters of home power systems are paying twice as much for their electricity as they would for conventional utility power (exclusive of line connection costs that would be paid for connecting homes in remote areas). Estimates outside the movement suggest that the cost penalty may be even larger.[10] In addition, they must pay essentially all of their electricity costs up front, unlike utilities that pay for fuels over extended periods of time, and utility customers that pay by the month as energy is used. Home power system costs, furthermore, cannot generally be included in a home mortgage.[11] They may even make it impossible to obtain a mortgage. High electricity supply costs quickly transmit the message to most participants in the home power movement that the most assiduous efforts to be efficient in the use of electricity will be the key to affordability for the home power approach as a whole.

As a product of heightened concerns for efficiency in the use of electricity, virtually every retail outlet for home power systems includes an array of superefficient lighting fixtures, lighting being perhaps the most essential use of electricity. These are displayed with their power demands in a way that allows visual comparison of light output from conventional incandescent and new superefficient fluorescent fixtures. Concern for efficiency does not end with lighting, however. A superefficient line of refrigerators and freezers has, for example, been developed and is primarily marketed to home power people. Assembled in a small plant in Arcata, California, by a company employing perhaps twenty-five people, Sunfrost refrigerators are the most efficient on the market in the United States, using on the order of one sixth of the electrical energy consumed by other commercially available models of comparable size; the 16 ft^3 Sunfrost refrigerator-freezer uses about 0.5 kWh/day. More efficient and reliable inverters to run 110 V ac appliances have also been developed in large part as a component of home power systems; roughly half the inverters sold by Trace Engineering (a pioneer in the small inverter business) are said to go to home power systems (the other half to boats, service vehicles, and recreation vehicles).[12] Highly efficient well pumps have also been developed for use with home power systems and optimized for minimal energy use with specific depth and flow-rate requirements.[13] Early in the movement's history, before inverter improvements began to take hold around 1990, dealer catalogues and retail stores offered 12 V appliances from televisions to room fans and kitchen blenders, all designed to avoid the energy losses and added cost associated with an inverter. Many of these were adapted from conventional 110 V appliances through a collection of cottage industries spread throughout the movement. In addition to these customized end-use devices, participants in the movement have developed and now produce and market an array of programmable and highly efficient battery charge controllers and other supply efficiency enhancers.[14]

Development of the Home Power Movement

The home power movement has its origins among individuals and small groups that moved to remote areas of California, Washington, and other states in the 1960s and 1970s. J. Schaeffer, for example, the current head and one of the founders of the largest mail order business selling home power equipment (the Real Goods Trading Company), moved to a remote part of California after finishing his degree in anthropology at Berkeley in the early 1970s. K. and R. Perez, early participants in the movement and publishers of its first real forum for exchange, *Home Power Magazine*, spent their meager savings on land on the California-Oregon border more than 10 miles from paved roads; moving there with a supply of food, they initially did not return to civilization at all for their first eight months. Another major mail order business (Backwoods Solar Electric) is run by S. and E. Willey, who also moved to their remote spot near Sandpoint, Idaho, in the early 1970s, living year around in an old school bus for their first five years.

In many cases, these were young people from the Viet Nam era, some of them broadly disillusioned with modern technology. W. Dankoff, an electrical engineering student at Carnegie Mellon between 1970 and 1972, for example, was discouraged from staying with engineering by a sense at the time that "technology [was] a more destructive force on the planet than constructive force." By the late 1970s, he had become a highly respected wind expert, primarily rebuilding old Jacobs wind generators. Since the drop in PV prices of late 1979, he has become one of the movement's leaders in developing highly efficient electric pumps for domestic water supply.

When these pioneers moved to remote locations, they typically made their move first and looked around for some way to make a living second;[15] by choice or necessity, they lived a meager existence without conventional homes, electricity, or even running water in many cases. Yet there was a desire more or less from the start for certain basic comforts, such as an electric light or the ability to enjoy recorded music. Small, inexpensive 12 V systems relying on automobile batteries began to appear, augmented in the more opulent cases by small gasoline generators freeing the battery from the car. In the middle to late 1970s, widespread attempts were made to improve on these simple systems by perfecting and applying wind generators to the battery-charging task. Real improvements in home power systems, however, became possible after 1979 or so, when ARCO Solar PV modules became affordably priced and eliminated the noisy, maintenance intensive gasoline generators or relegated them to a limited backup function.

The home power businesses now responsible for the vast bulk of home power sales emerged from their founders' personal efforts to develop minimal electric power supplies for their own remote homes. Working with their neighbors, assisting new arrivals, and often coordinating collective purchases of PV modules to take advantage of volume discounts, they gradually found themselves operating their own small businesses. Insufficient to support their founders for some years, these businesses grew largely without loans or venture capital of any kind, sustained initially through unrelated part time or full time jobs. Over time, small home power communities, sometimes linked by

CB and radio telephone, developed around and among the early pioneers. As relatively isolated home owners established better contact with each other, a nationwide community of home power enthusiasts began to emerge. By the summer of 1989, significant participation was beginning to be evident from increasingly diverse social groups including, for example, retirees wanting to move permanently to previously unimproved vacation homes.[16]

From a strict business standpoint, home power had begun to take on a higher profile by the summer of 1989. While the bulk of the business continued to be mail order from remotely located pioneers, new entrants were beginning to be more visible in retail stores. These new outlets often did much of their business in bicycles, woodstoves, PV-powered cattle watering systems[17] or other related markets, but also included large floor area and stock commitments to home power systems. At least one multistore conglomerate (Photocomm) had emerged through the acquisition of several independent businesses.

Participation and Motivations

There are probably more home power systems in California than in any other state and there appear to be more in the West than in the East,[18] but significant participation in the movement could probably be unearthed in every state in the United States.[19] It now spans a broad range of incomes and patterns of life from the most frugal to those that include a satellite dish and VCR. Participants are generally college educated and come from middle class backgrounds,[20] yet they widely appear to share strongly felt motives that have carried them far from traditional patterns of life.

Participants in the home power movement almost uniformly have a clear desire to live differently. In some cases they appear to have made their choices based on inchoate or relatively unexamined drives, in others, on clearly articulated social and philosophical positions they developed over many years. In either case, there is clear and recurring evidence of a desire to reshape the human relationships of community and traditional definitions of work roles and a desire to redefine relationships with the natural world. Yet this is not a utopian movement. Self-sufficiency, for example, is pursued only to a modest degree and as a means to assure the independence necessary to a reshaping of patterns of life in the ways just mentioned.

The desire for a strengthened sense of community is perhaps most evident in the five actual communities of home power adopters the author visited in the summer of 1989 in the states of Idaho, Washington, Oregon, and California.[21] With between 10 and 100 homes in each of these communities, everyone in each case apparently knew virtually everyone else in their community well, sometimes having come to the area with other members of the community. An unusually high degree of internal cooperation was evident in the "home schooling" systems set up in several of these communities (partly necessitated by remoteness from regular public schools), in the frequency of cooperative home construction, road maintenance, and other projects, and in the way each community seems to rely on a kind of internal specialization. The development of home power businesses, as described, exemplifies this kind of specialization as it has overflowed beyond the communities in which it began. Other observed community service specialities include CB and radio telephone communications systems, ram pump and other water supply systems, and home schooling teachers. Members of one community had even cooperated in a kind of police function to end marijuana growing; water had been supplied free to a newcomer by a neighbor with an oversized system, but when use of that water to irrigate marijuana plants became known, the water supply was cut off and the newcomer was strongly encouraged to leave. In a few cases, these communities have now become sufficiently densely settled that conventional electric utility service has become an option; in these few cases, the switch to traditional utility service has generally been rejected by the communities themselves.

Efforts to redefine work roles have been significant both in terms of human relations and in terms of work content. Well over half the participants interviewed by the author in the summer of 1989[22] had built or participated in the construction of their own homes. All of them were closely involved in the design, installation, and operation of their own home power systems. Many apparently preferred a less specialized, more diverse mix of work activities, combining part time or seasonal paid work with self-employment in forms such as cutting their own fuel wood or growing some of their own food. Although most of the people in the home power business had by 1989 become busy on more than a full-time basis, this preference for an innovative mix of work structures and activities even extended in one case to the home power business itself: only one out of the ten or so employees of Alternative Energy Engineering, a major retailer in Garberville, California, chose to work full time in the business.

Environmental values are among the most commonly expressed explicit motivations for participation in the home power movement. "Mr. X," for example, first built his own PV-powered log house while working on his degree in forestry, then started a business selling home energy systems and PV-powered remote livestock watering systems in western Montana. Grandson of one of the original opponents of clear cutting in the Bitteroot forests and son of the head of a Washington-based environmental organization, X had decided, upon completing his degree, that the most effective way to address his own environmental concerns would be through the development of alternative energy systems. He worked for two years for Solarex in Maryland (one of the largest U.S. producers of PV panels), then started his present business back in Montana with a $4,000 loan from a close relative. Over a period of three years, he and his wife managed to develop a viable business and purchased a house in a neighborhood in town which they converted to a roof-mounted PV system (about 500 W peak). As of 1989, their newly rented storefront remained a hub of activity for the grassroots organization Friends of the Bitteroot, as well as for home power efforts in their area.

Environmental commitments are, of course, widely evident in participants' frequent choice of beautiful remote locations for their homes. Activism on local environmental issues is common also. In addition to X's activities on behalf of the Bitteroot forests, other participants have alerted officials to the presence of bald eagles in proposed lumber harvest areas and protested overgrazing in federal permit areas near their homes. In a few cases where systems have been installed in homes only a few feet from utility lines, environmental concerns seem to have been an important component of an almost religious set of commitments to a new pattern of life.

In a more general sense, home power systems have been adopted as part of an effort to reduce living expenses as a further means to increased freedom of action. This theme is perhaps best illustrated through the details of Y's experience in the woods of central Minnesota. Y built his approximately 1,000 square foot underground home himself in 1983, partly from timber on his land. It is fully paid for: he reports $2,500 for the land and $8,000 for purchased building materials. Y has two small PV panels, a hand-pumped shallow well water supply, wood heat, and a gas stove. He works intermittently building log cabins and in the lumber industry but he has made a deliberate attempt to reduce his living expenses to a minimum. Although he is only about 200 feet from a utility line, he argues that connections of this sort make people serfs of the system: once entangled in these and other "tendrils" (the mortgage on a traditional home would be another example), people are compelled to spend their time in ways that allow them to earn the money to pay for the services provided. Y spends about three months each year on extended trips, a practice made possible by the very low cost of living associated with his home power home. When he was interviewed in 1989 he had just returned from Nicaragua where he had been working as a volunteer on a portable sawmill project for more than a month. Y's politics were shaped in part by his two years in the service in Viet Nam; he reported a conversation with a CIA pilot that left him firmly convinced that the CIA was transporting drugs to help finance the secret war in neighboring countries, drugs that would eventually be purchased in part by U.S. servicemen. Y is affable and easy going despite his somewhat radical politics, perhaps because he has managed to articulate the origins of his discomfort with traditional policies and because he is comfortable with the innovative pattern of life he has implemented for himself. While Y's politics are not necessarily representative of the home power movement, his effort to minimize living expenses partly as a means of freeing his own time to pursue activities of his own choice *does* seem to be typical of a large segment of the movement.

Interpretation and Significance

The behavior of those involved in the home power movement lies well beyond traditional images of technical or economic rationality. It is true that the vast majority of home power systems have been installed in remotely located homes where utility connection charges would amount to a large fraction of home power system costs. It is also true that remote siting has generally meant relatively low land costs. In this sense, land, home, and power costs taken together may be lower, in many home power situations, than comparable conventional land, home, and power costs. This fact, widely noted as an explanation for their actions by home power people themselves is, however, insufficient to explain either the remote location decision itself or the other attributes of the home power choice. The decision to live in remote locations, alone, generally implies severe restrictions on employment opportunities and severe income penalties as compared with more densely populated areas.

The home power decision is eminently rational, on the other hand, in the sense that home power people are generally well satisfied that they have gotten the most of what is valuable to them in exchange for their resource expenditures. As a product of the home power "experiment," this result is of great importance in raising a more concrete challenge to prevailing technical and economic approaches. Focused specifically on relative electricity costs, the traditional methods fail to predict or comprehend the kinds of choices participants in the home power movement have made. In the process, traditional methods have failed to recognize, let alone encourage, developments with the potential for significant resource conservation, environmental, and other social benefits, and effectively screen this and perhaps other promising alternatives from broadly based social consideration. In concrete terms, this failure helps to reveal the narrow "upward mobile middle class"[23] value assumptions that are implicit in traditional energy policy analyses.

In practical terms, participants in the home power movement may also have contributed significantly to the development and demonstration of alternative patterns of life that could help to resolve many of the environmental problems associated with energy production and use (i.e., the problems of global warming, acid deposition, oil spills, radioactive wastes, and so on). Environmental benefits accrue both from the generally less damaging renewable energy systems themselves,[24] and, in all probability, from the marked reduction in energy use associated with efficiency improvements much larger than those that would be economically justified with conventional energy sources now available at lower market prices. Even at its present level of development, the home power option poses an interesting challenge, both to individuals and to the policy community concerned with environmental issues. Given its higher price, particularly in grid-connected locations, and its environmental benefits, the home power alternative raises the challenge, "How much are environmental improvements worth to you?" in quite graphic terms. Home power and perhaps other environmentally attractive departures from traditional sociotechnical expectations[25] may need to be seriously examined, even at higher prices.

Participants in the home power movement have discovered for themselves that more expensive (per kWh) renewable home power sources, efficiency improvements, and end-use adjustments can be arranged in affordable combinations that are more attractive than conventional energy systems in the context of commitments to the environment and a restructuring of work content and human-community relations. More expensive power supplies and lighting and home appliance systems have, in fact, been used in many cases to *reduce* the total cost of energy use in absolute terms as a further means of serving those same commitments. In an odd reversal of the usual patterns of consumer behavior, home power adopters seem to have internalized nearly all of the social and environmental effects of their energy choices in their energy decision making. In a further departure from widespread consumer behavior, home power decisions appear to be biased in ways that cause more energy conservation than would be expected from an economically rational response to price.[26] The analytical methods employed by home power adopters may remain, in many cases, less quantitatively sophisticated than traditional methods, but it can be argued that the adopters' decision processes are substantially more sophisticated than traditional methods with respect to the integration of disparate factors into coherent choices. The home power movement suggests in a small way that the assumption that consumers act only from very narrowly conceived self-interest (and imperfectly at that) may be misplaced. The capacity for choice based on more broadly conceived self-interest, and for integrative conceptualization of more sustainable patterns of life based on those broadened concepts of self interest may, this movement would suggest, have been underestimated.

Not all such experimental engagement with the realm of possibility will, of course, be so successful either by the standards of participants or by external standards such as levels of environmental degradation. Standards of "efficiency" will necessarily have to be adjusted in keeping with this fact. Efficiency in the development of new patterns of life is something quite different from efficiency in mass production or in attempts at the perpetuation of a fixed pattern of life. In this respect perhaps more than any other, the home power movement is remarkable. Beginning without any clear notion of where they would end up, participants have progressively developed significantly different patterns of life that now appear to be reasonably functional. In the course of directly engaging all aspects of their energy choices, they appear to have succeeded in arriving at "solutions" that are more consistently expressive of their own values and preferences than traditional patterns have been. And they have accomplished this without the massive infusions of social resources that have backed the expression of more widely assumed preferences—infusions that might be regarded as extortionist by those who do not share the same preferences.

It is extraordinarily difficult to "think" one's way to new and preferable patterns of life accompanied by new and preferable systems for energy production and use. It is remarkably difficult even to sort out how present energy patterns shape the way we live now. In the face of such difficulties, experimental approaches that build almost blindly from initial responses to energy or living stresses may prove the only practicable way to arrive at improvements. Without abandoning what can be gained from theory and abstract thought, such a conclusion would suggest that we need to end our elaborate avoidance of the issues and get about the business at hand.

A Theoretical Framework for Home Power

It is a common practice among renewable energy enthusiasts to collect "success stories"[27] illustrating the widespread feasibility of renewable applications in a world that generally regards them as impractical or uneconomical. It is important to note, however, that the home power movement is not just another success story in the normal sense. When we extol the virtues of wind power production and point to the more than 1000 MW of installed wind capacity in California or when we celebrate innumerable local cooperative achievements in implementing solar hot water heating systems, solar greenhouses, and so on, we largely celebrate situations in which a renewable energy device has been substituted for conventional energy production with little other change. The large wind systems in California, for example, are grid connected and entirely invisible to utility customers who do not happen to live in or drive through the mountain passes where the machines are sited. Our usual definition of "success stories," in fact, accepts uncritically the traditional measures of "success," including economic profitability and social and technical compatability with existing institutions, and does little to express a rethinking of how we live our lives.

Although the usual litany of renewable success stories may raise the spirits of renewable energy enthusiasts and may be useful in prodding others to think somewhat more broadly, it ultimately does little to encourage us to address the fundamental energy related questions outlined at the beginning of this book. By implicitly endorsing traditional standards of success, in fact, it may have the opposite effect. More than is recognized may actually be going on in much of the renewable energy activity that we tout as a success by traditional standards. But our acceptance of traditional standards of success in arguing its significance tends to undermine the kinds of thinking this book urges as necessary and worthwhile.[28]

In the context of this book, the unique value of the home power movement as distinct from other success stories is that it retains at its core an unmistakable departure from the way we traditionally live our lives. It is categorically different, even from vigorous programs such as that of the Sacramento Municipal Utility District (SMUD)[29] for placing photovoltaic systems on single family dwellings. It steps outside the normal frameworks within which *success* has been defined. It does not "succeed" simply in replacing one machine for producing electricity with another. Instead, it alters fundamentally the relationship between the energy user and the energy production process. It has necessitated intimate participation by the homeowner in thinking about each household use of energy. And it has developed as an expression of fundamental reformulations of community, work, and environmental relationships. In examining the home power movement, it is not possible to see the energy production systems that have been adopted as simple replacements for more traditional sources.

In trying to grasp the uniqueness of home power activity, it may be useful to expand substantially on our earlier summary of the conceptual framework offered by Albert Borgmann's work in the philosophy of technology. As a departure from customary practice, home power activity may seem unexceptional when we understand it in terms of localized commitments to community, environmental, and work-related values. Placing the movement within an alternative theoretical framework, on the other hand, highlights its potential significance not as an exception posing no challenge to the rules, but as an illustration of behavior consistent with a new set of rules. The success of home power, that is, may not be simply the success of having located a "loophole" permitting innovative practices within customary boundaries. Its success may lie instead in the definition and adoption of a new set of rules altogether.

Useful insights into the home power movement can surely be culled from a number of bodies of theory including the study of the diffusion of innovations[30] and the study of social movements and collective behavior.[31] Albert Borgmann's work in the philosophy of technology, however, appears to offer uniquely useful insights, placing the movement in the context of a general characterization of contemporary technological society. To see this, a more extended summary of Borgmann's work will be required. We can begin with a more detailed summary of his use of the terms "thing" and "device," and eventually apply the framework they establish to our thinking about home power.[32]

A *thing*, in Borgmann's use of the term, brings with it bodily and/or social "engagement" with the thing's world. In this sense a thing necessarily brings with it more than any single commodity it may make available. Thus, in many cases, a wood stove once

> used to furnish more than mere warmth. It was a *focus*, a hearth, a place that gathered the work and leisure of a family and gave the house a center. Its coldness marked the morning and the spreading of its warmth the beginning of the day. It assigned to the different family members tasks that defined their place in the household. The mother built the fire, the children kept the firebox filled, and the father cut the firewood. It provided for the entire family a regular and bodily engagement with the rhythm of the seasons that was woven together of the threat of cold and the solace of warmth, the smell of wood smoke, the exertion of sawing and of carrying, the teaching of skills, and the fidelity to daily tasks. These features of physical engagement and of family relations are only first indications of the full dimensions of a thing's world.[33]

In Borgmann's typology, a *device*, by contrast, serves to make a single commodity highly available while making the mechanism of procurement recede from view. Thus a furnace ordinarily provides mere warmth (its "commodity"), preferably in an instantaneous, ubiquitous, safe, and easy way. The "device," then, disburdens us of both social and bodily engagement, leaving only the commodity (warmth) in evidence.[34] "A device such as a central heating plant procures mere warmth and disburdens us of all other elements. These are taken over by the machinery of the device. The machinery makes no demands on our skill, strength, or attention, and it is less demanding the less it makes its presence felt. In the progress of technology, the machinery of a device has therefore a tendency to become concealed or to shrink.

Of all the physical properties of a device, those alone are crucial and prominent which constitute the commodity that the device procures."[35] As a characteristic of contemporary life, Borgmann suggests, we are inclined to think of the physical and social engagement of "things" (e.g., woodstoves) as burdensome. And, in the effort to disburden ourselves, we move increasingly from a world of things into a contemporary world of devices. Under such a "device paradigm," we move away from social and bodily engagement, away from skill, and toward increasing commodity availability, deskilling, anonymity, and disengagement.

Many instances of engagement are unquestionably experienced as burdensome. To be clear, Borgmann's intent is not necessarily to advocate a return to wood stoves, nor is it to urge uniform avoidance of disburdening technology. For some, woodstoves may be associated with oppressive class relations under which it would be the servants that would be up early to start the stove and up late as well to bank the fires. Or they may be associated with a backward and determinedly uneducated way of life, abusive of wives and children, and decidedly not consistent with values we would seek to serve. Indeed the distinction between things and devices need not lie entirely in the technology itself. An old furnace that is known, coaxed, nursed, and repaired might be as much a thing as the wood stove of Borgmann's description. An automobile or a bicycle could in theory be a thing or a device depending upon one's relationship with each of these technologies. As Cowan[36] and others have noted, our relationships, even with the technologies of the home, can be quite complex in practice. The point here lies in the nature of the relationship and in the nature of the choice to adopt and use a technology, not in any unequivocal categorization of particular technologies, whether woodstoves, furnaces, cars, or otherwise.

Disburdenment, in any case, is not the only significant feature of the move toward devices. A further illustration of the losses that may, in certain instances, be involved, is provided by a cartoon Borgmann describes in which a middle-aged woman stands in front of a chest of frozen dinners in a supermarket. . . . [H]olding up two packages, looking a little puzzled . . . she says to her husband: For the big day, Harv, which do you want? The traditional American Christmas turkey dinner with mashed potatoes, giblet gravy, oyster dressing, cranberry sauce and tiny green peas or the old English Christmas goose dinner with chestnut stuffing, boiled potatoes, brussels sprout and plum pudding?" Harvey looks skeptical and a bit morose. The world of bountiful harvests, careful preparations, and festive meals has become a faint and ironical echo. Mabel is asking Harvey whether on December 25th he would rather consume this aggregate of commodities or that. To consume is to use up an isolated entity without preparation, resonance, and consequence. What half dawns on Mabel and Harvey is the equivocation in calling the content of an aluminum package a "traditional American Christmas turkey dinner." The content, even when warmed and served, is a sharply reduced aspect of the once full-bodied affair."[39]

While the various preparation and preservation mechanisms hidden from view by the device providing frozen Christmas dinners have, in fact, removed the social and bodily burdens of preparing a Christmas dinner in the traditional way,[38] the cartoonist here surely taps a widely shared sense that *this* disburdenment, at least, is a mixed blessing.

Any technology necessarily incorporates certain elements of action and fixes certain social and physical relationships. In the process, it places constraints on action and on the development and expression of values through action. Where the implications of those constraints for the expression of values are taken to be inconsequential compared to the commodities procured by the devices involved, the desired expression of values may progressively be substantially frustrated.

The design, implementation, and operation of home power systems suggests that their selection is not consistent with the device paradigm. The device (utility power available at the flip of a switch) has been rejected in favor of a more complicated and more expensive (more burdensome) technology from the user's standpoint. In choosing to make use of home power technologies, participants in the movement have effectively sought to reclaim expressive elements of action that have otherwise remained fixed in the mechanisms of traditional devices including those employed in providing electricity as a commodity. In their efforts to live differently, movement participants have rejected the usual device approach, returning instead to a certain engagement with things as a means of reasserting values or commitments in their own social and environmental existence that had been frustrated under the traditional device approach.

Borgmann places emphasis on what he calls "focal concerns" in his proposals for a reform of technological patterns of life dominated by the device paradigm. As an alternative to the predominant, relatively uncritical embrace of devices, he argues the need for a more direct consideration of what might constitute the good life. He suggests that we consider orienting our decisions about technology not simply around an unexamined disburdening process, but around what he calls "focal things" and "focal practices." If one's notion of the good life, for example, is sustained by the bodily and social engagement of the wood stove, then it should perhaps be retained. The wood stove could, in other words, be regarded as a focal thing, and the associated exercise of physical skills and social relationships could be regarded as focal practices in a particular definition of the good life. Under such a definition, it would be a mistake to replace the wood stove with a modern oil-fired furnace.

Borgmann suggests that there may be evidence of the reforms he advocates in the popularity of running as an avocation, in the appeal of careful food preparation, and in other practices in which focal concerns have outweighed the traditional device approaches (in these cases, outweighing "efficient" transportation and instantaneous meal preparation). In each case, focal practices have been developed as a way of asserting, defending, and expressing focal concerns. Borgmann's proposals in this area could well have been used as a guiding theory in conceptualizing the solutions of the home power movement. The focal practices and concerns of community, work, and environmental relations, and the unique combinations of technology (focal things) they have elicited, form a clear foundation for the progress of the home power movement from the perspective of those involved. The reinvigorated expression of participants' values and the reform of patterns of life in the direction of what Borgmann refers to as "the good life," (from the perspective of those involved) has been a clear product of giving these focal concerns a central place in the lives of participants, very much as Borgmann[39] recommends. Home power can be seen to rest on a "practice of engagement." And it can be seen to arise from a concern with focal things and focal practices in the sense that the renewable energy technologies employed and their practical utilization in the context of community and the natural environment, provide a "center of orientation" for participants in the movement.

At a more detailed level, Borgmann's distinction between "wealth" and "affluence," echoing Thoreau and other earlier sources, could well have been used as an explicit guide in the practical case of the home power movement.

> There is prosperity . . . in knowing that one is able to engage in a focal practice. . . . one prospers not only in being engaged in a profound and living center but also in having a view of the world at large in its essential political, cultural, and scientific dimensions. Such a life is centrally prosperous, of course, in opening up a familiar world where things stand out clearly and steadily, where life has a rhythm and depth, where we encounter our fellow human beings in the fullness of their capacities, and where we know ourselves to be equal to that world in depth and strength.
>
> This kind of prosperity is made possible by technology, and it is centered in a focal concern. Let us call it wealth to distinguish it from the prosperity that is confined to technology and that I want to call affluence. Affluence consists in the possession and consumption of the most numerous, refined, and varied commodities.[40]

Borgmann argues that "it must be recognized all along that a reform of technology will diminish affluence but increase wealth."[41] But he notes that "One who is engaged in a focal practice . . . can reduce consumption without resentment."[42] These insights undeniably appear in concrete translations throughout the home power movement as participants have consciously chosen reduced material affluence in favor of increased wealth and continue years later to be well satisfied with the choice.

In the home power context, Borgmann's theory proves well worth examining as a practical guide in efforts to escape the device paradigm. In drawing this link, it is important to note that neither the home power movement nor Borgmann's theory calls for a retreat from technology as a part of efforts to break with the device paradigm. Borgmann has explicitly indicated that wealth "is made possible by technology." The point instead is that wealth and affluence may only be achievable through distinctive selections of technology. Borgmann's emphasis on focal things and practices as the key to reform may, again judging from the home power experience, be particularly efficacious.

Framed in this way, home power becomes not merely a localized exception to the rule, a "success" in the sense of artful exploitation of a loophole, but a decision to be guided by a different set of rules. The renewable energy systems being adopted are seen not as locally suitable substitute suppliers of kilowatt hours, but as an integral part of broad commitments to a particular way of life. Judging from the home power movement, Borgmann's suggestion that we choose technologies in relation to focal things and practices may be a valuable guide in the exploration of energy alternatives for the future. Such a direct approach to the question of what might constitute the good life at least has the great advantage of loosening traditional limitations on the alternatives we are allowed to consider.

Guidelines for Explorers

From a substantive standpoint, the home power movement may be no more than a passing curiosity for many readers. Whatever its substantive value, however, the movement is most important in the context of this book as a demonstration of the kind of exploration and experimentation that appears to be necessary if we are to appreciate fully the range of possibility, both technical and sociocultural, that energy options afford. Given the concerns of previous chapters with regard both to existing energy practices and to the theoretical constructs usually employed in guiding energy choices, it is important that we somehow get beyond the constraints imposed by traditional theory and practice. This is not to say, for instance, that we should simply discard traditional economics. But we need to move to a higher plane of cultural and historic possibility than is attainable within a strict adherence to traditional economic and other standards. Exploring the "option space" in the way that home power enthusiasts have may be a very effective way to make this move. If we are truly to consider more than one candidate in choosing our energy future, we will need to suspend even the more reliable guides of the past and *experiment*.

Beyond this central message, the home power movement also raises some interesting questions about possible guides to the process of experimentation and exploration. Is it significant, for example, that the home power movement has been very much a "tinkerers' movement"? The energy supply technologies employed have almost uniformly been relatively easily understood and manipulated by ordinary individuals. And this manipulation has been possible in real settings in direct contact with both a set of equally accessible energy using technologies and the requirements and constraints of life in the settings in which they were to be employed. This is a marked contrast with something like nuclear power, which has perhaps suffered from its development by highly specialized experts relatively isolated from the social environment in which the technology was ultimately to be implemented. If energy systems are to serve human needs and be expressive of the values held by those who adopt them, it may be important to preserve the ability to tinker with them in their development. At very least, it may be useful to bring experts and ordinary people into closer proximity so that they can "tinker" collectively, avoiding the underexamined definitions of the energy problem that seem to have undermined past solutions.[43]

Energy and other policies that have been governed for some decades by centralized decision making may also have been constrained in ways that the individual actions of home power enthusiasts have not been constrained. If a government body is to fund research on a particular energy option, it can legitimately do so only on the basis of criteria that enjoy broad support in the society as a whole. In recent years, this has tended to imply that support could only be channeled to alternatives that showed strong promise of being cost effective by traditionally accepted standards. Alternatives such as those pursued in the home power movement remained beyond the acceptable range. But the process of policy development and the consideration of energy alternatives may have become, as a result, a predominantly "instrumental" process, constraining the evolution of decision criteria in spite of substantial shifts in resource, environmental, social, and other relevant conditions.

Decision making can, after all, be as much a "constitutive" process as it is instrumental.[44] Our decision making and the experience that flows from it can, in other words, be as important in helping us to define who we are and what our goals and objectives are as they are in actually bringing us closer to achieving those goals and objectives. If we limit ourselves to choices that we are fairly sure will lead to success by preexisting standards, we may displace decisions that we are unable to direct toward specific objectives but that could nevertheless have been productive in helping us to formulate and define what is and is not desirable.

Home power enthusiasts, because they have been working almost entirely from their own resources, have been in a position to experiment more freely, make choices with implications they could not clearly define, and learn lessons not accessible within a strictly instrumental context.

If we are to revive the constitutive and deemphasize the instrumental aspect of our energy decision making, experimentation such as that undertaken in the home power movement may, again, be very useful. Incentives for experts from all disciplines to work more closely together and to work collectively more closely with ordinary people may be valuable. Criteria for the allocation of public resources may also need to be changed. Less public funding for the development of specific energy alternatives might be appropriate to provide more room, relatively speaking, for the emergence of unconventional alternatives. Looser funding aimed less at new technologies sure to fit old criteria and more at social and technical experimentation designed to pursue questions about those criteria may be in order. Ideally, the latter would be leveraged against personal commitments to nascent alternative perspectives. Personal commitment, after all, has carried and sustained the home power movement; commitment beyond the profit motive of a federal grant is likely to be particularly important as the pursuit of new alternatives inevitably encounters complications not imagined at the outset. Again, it may be appropriate to anticipate a very high "failure rate" as a high proportion of experiments lead to apparent dead ends. We must remember that much can be learned from failure. Columbus did not discover a new westward passage to the Far East; yet his "failure" led to material and cultural developments of historic proportions.

In closing this chapter it is perhaps worth emphasizing that I do not mean to encourage romantic pursuits that are clearly materially infeasible. A home biogas plant operated from human wastes alone, for example, simply cannot produce enough energy to supply ordinary cooking, lighting, and other household needs. "Pedal power" cannot simply be substituted for the diesel engine beneath the hood of a trailer truck. Nor do I mean to urge the pursuit of radical social change where that change would be achieved through coercion or even through substantial "incentives" (which tend, in my view, to encourage us to behave like sheep and undermine any inclination we might otherwise have had to engage our problems directly, ourselves). It might well be interesting to look at biogas plants of some kind as a source of backup fuels to augment storage systems associated with intermittent renewable energy supplies. It may also be interesting to consider means more efficient than diesel trucks for moving freight—as well as patterns of production and use that might reduce the need to move freight in the first place. And we might do well at least to recognize and provide for the possibility of significant social changes either at a localized level or in society as a whole. What these require, however, is not abandonment of, but closer attention to, our standards of material possibility and of individual and collective desirability.

It is perhaps worth reiterating the point also that the central purpose here is not to urge attention to the home power movement as The Solution to the energy problem. Home power may or may not prove to be of substantive interest to individual readers or to significant segments of society as a whole in the long run. Its spread and increasing visibility may or may not elicit changes over time in the expression of preferences among those now uninterested in anything beyond the availabiltiy of power, whatever the source, at the flip of a switch. Present home power technologies may or may not prove attractive and affordable in grid-connected applications or as an element in urban energy futures characterized by vastly increased end use efficiencies. Without doubt many interesting issues remain to be pursued here. Personally, I do advocate pursuing them. But ultimately, these are not the central issues.

The central value of an examination of the home power movement lies in its unique usefulness in gaining access to a very different perspective on energy issues—a usefulness much enhanced by its proximity to our own experience (i.e., we are not describing the energy practices of Australian aborigines here) and by the unusual degree to which concrete expression of that perspective has been achieved in practice. In the context of this book, the principal value of the movement lies in the concrete support it provides for the otherwise abstract notion that conventional treatments of energy issues may be expressive of a radically impoverished imagination, carving an impossibly narrow collection of options from a much more expansive terrain of technical and sociocultural possibility. On material grounds, out of respect for human diversity, and in the interest of discovering paths for the future that will prove engaging and satisfying, as well as materially sustainable, the purpose here is to urge a much more aggressive exploration of that expansive terrain of possibility and a more conscientious effort to engage the human imagination, before we settle into any solution or solutions for the future.

SEVEN

Where Does This Leave Us?

Anyone with the observer's instincts of the social scientist[1] will likely remain intrigued by our collective response to energy issues. The sweeping generalizations and bold assertions of previous chapters notwithstanding, that response will inevitably remain enigmatic and mysterious at some level for all of us. How is it, we might ask even now, that the same period of ten or twenty years has seen both the widespread adoption of personal computers in the home and the failure of home power systems even to achieve broad recognition as an alternative for consideration? Each of these new technologies involves roughly the same incremental cost, and neither initially offered immediate materially palpable benefits to the individual adopter.[2] Who has chosen the directions implicit in our energy decisions? How have we arrived at this particular juncture?

I will not review in detail here the attempts to answer such questions that have been offered in earlier chapters; to list them, we have spoken of institutional and collective momentum, individual or popular nonparticipation, particular distributions of established interests, and a body of established theory and practice consistent with and supportive of each of these. We also need not dwell on the notion that our present situation is a result necessitated neither by "the facts" nor by definitive theoretical analysis. The case has been made, as best I can make it, that present energy policies and technologies, and the pat-

113

terns of life they support, are profoundly *underdetermined* by existing theory and fact, and that they are instead very much *socially constructed*, the usual trappings of scientific analysis notwithstanding.[3] It has further been suggested that present policy and practice has been selected from a range both of technical and of sociocultural possibility that is vast when compared with the alternatives admitted for consideration under commonly adopted notions of "feasibility." Contrasts with developments outside the energy sector, with response, for example, to technologies like the home computer, provide only stark confirmation of these conclusions.

Rather than reviewing in any greater detail what has been said already, this chapter is devoted to thoughts about where what has been said leaves us and how we might wish to proceed from here.

Lost?

Are we lost then? If we accept the indeterminacy of the situation and agree to rethink everything from scratch and engage our creative imaginations in a truly vast option space, where does this leave us if not simply lost? Where are we exactly, and where should we be trying to go?

Thinking and experimentation of the sort urged in this book can, in themselves, be very unsettling and even threatening in the absence of foundations at some deeper level. To the degree that our sense of security is anchored at a relatively superficial level in traditional patterns of life, depending, for example, on "two cars in every garage" or on an accustomed progression of new technologies destined to culminate in our eventual control over nature, we may be in serious trouble. Those whose sense of security rests in an altogether nonmaterial plane or requires no more than a guarantee of the next meal and a warm place to sleep, on the other hand, may be in a position to exercise a wider range of imagination in thinking about energy futures. Ultimately, actual energy preferences may be strongly related to such notions of security. Varying with personal background those preferences may be strongly affected by individual perceptions of the distance between the resources now available to us and the "necessities" of life and family as we see them.

There are, however, a number of ways we might set up meaningful experiments. In general terms, we can begin either from the human side, as the home power movement has to a degree, or from the technical side. In human terms, experiments can be initiated by listening more broadly and more carefully to the diversity of cultural threads ordinarily drowned out by mass culture. From one perspective or another, there will always be dissatisfactions and at least inchoate desires for improvement in the conditions of life. As in the case of the home power movement, these are available to be tapped and pursued in new directions if we listen carefully enough for them and throw our net wide enough in society (i.e., beyond the relatively narrow set of energy analysts committed to "upward mobile middle class values") to capture them. Mechanistically, a wide range of experiments could be built from the ethnographer's approach to understanding various groups and subcultures in the United States today.[4]

Beginning from the technical side, experiments can be developed by returning to the most basic physical understandings of what is possible materially. One might look, for example, at energy flows in the transportation sector. Oil use accounts for about 40 percent of U.S. energy use and two thirds of that goes to transportation. U.S. oil production peaked in 1970, and world production is expected to peak by around the year 2000.[5] In addition, the only relatively easy substitute for oil in transportation, natural gas, is in roughly as short supply as oil in the long run. Proceeding purely from this technical description, we might be prompted to examine communities arranged to eliminate the need for cars in daily shopping and commuting. Rather than trying to find an electric substitute for the present auto—a solution exemplary of the kinds of institutional momentum discussed in earlier chapters—we might look again at walking, enhanced urban mass transit, and extremely inexpensive, very short range, light weight, low speed vehicles to assist with occasional loads like groceries. Technically speaking, such arrangements, in combination with long distance mass transit, could sharply reduce transportation energy use. People once got along reasonably well without automobiles; given the energy situation, perhaps it would be worth reexamining the implications, both positive and negative, of cutting back sharply on the present use of this technology.

In coming up with ideas for experiments, it will be very important to ease traditional notions of "feasibility" and to distinguish more carefully between actual physical possibility and social or political feasibility. A certain playfulness may need to be cultivated. As in the practice of brainstorming, it will be important not to eliminate ideas in their early stages of formulation simply because they do not appear to be economically practical in the usual narrow sense. Similarly, possibilities suggested by technical consideration of energy flows may be worthy of consideration even if they might imply significant changes in social practice. If we overcome our initial reluctance, we may find that many such changes would prove welcome in ways unrelated to their narrowly defined energy benefits.

As aids to the imagination, quasi-formal procedures that might be used in generating experiments are beginning to emerge from a growing concern with democratic control of technology. These range from mechanisms designed to enhance public participation,[6] to institutional and intellectual frameworks for "constructive technology assessment."[7] In the latter, attention is given to mechanisms for "stimulating alternative variations, changing the selection environment, and creating or utilizing technological nexus."[8] (The last of these focuses on alternative links that may be created between the variation process and the selection process.)

To take a single, relatively technical example that might arise in a constructive technology assessment framework, the selection environment for refrigerator design might be changed by shifting the design problem from the manufacturing and marketing context of General Electric, for example, to an undergraduate engineering class. Instructed to design a refrigerator that would use as little energy as possible, students would likely arrive at a whole array of alternative variations. Vastly increased insulation thickness or perhaps even evacuated panel insulation (like a thermos bottle) might be introduced. The compressor, which gets hot, might be moved from its usual location beneath the chilled space (warm air rises) to above it, or even to another room. Students might even consider placing the compressor and condenser coils (visible on the back of some refrigerators) outdoors like the analogous heat rejection components of an air conditioning system, so that the refrigerator could take advantage of winter temperatures that would much reduce its cooling task. Although appliance designs along this line are not now mass produced, such an exercise in constructive technology assessment might lead to interesting images of alternative ground rules and patterns of life in which a redesigned refrigerator of this sort might be appropriate. (As it turns out, all but the last of the design changes described are among those incorporated in the Sunfrost units produced for use in the home power movement.)

It is beyond the scope of this book, and beyond the capacity and proper role of one or a few minds, to try to offer a definitive list of appropriate experiments to be undertaken. If we are to judge from the variations that have appeared even under the most modest easing of constraints (see, for example, discussion of responses to the Public Utility Regulatory Policies Act in Chapter 5), the range of possibility is truly staggering. The point, in answer to the question that opened this section, is simply that, in spite of possible appearances to the contrary, we are not "lost." Within the constraints of our own personal sense of security, which may themselves be open to examination and adjustment, we all arrive within a context. We all arrive, at whatever age, with a lifetime of experience. We all are equipped with a certain capacity to examine that experience and to draw lessons from it. We all have some basic set of notions of how the world works and what our position is in it, of what is desirable and what is not. It may well be that we will need to go back a bit, suspend belief in some of the lessons we thought we had learned, and reexamine the past in light of new developments. But carefully considered, these foundations—our individual and collective experience and our native capabilities for finding order and meaning in that experience, augmented, perhaps by the aids to imagination of frameworks like constructive technology assessment—should prove an adequate guide. The archaeological record would at least suggest that such basic tools have stood the test of time to date.

How Do We Proceed?

If we are not lost, still, how are we to proceed? The best and only real answer to this question may be "one step at a time." But let me urge several other guidelines or directions for the process. First, as we continue to make choices affecting our future way of life, whether they are made in small increments or in larger leaps, we are going to need to work harder to improve communications among people holding differing points of view. Second, recognizing that fundamental differences are likely, we must try to minimize the unredeemable losses implied by the conflicts that will result. And, finally, we must consider the possibility that material outcomes may be less predictable, less knowable, and ultimately less important than how we treat each other in the course of our efforts to choose and effect a desirable future.

Better communications are essential if we are to understand each other. And understanding each other is the only way to avoid getting in each other's way unnecessarily. Understanding is fundamental as a basis for handling differences fairly. If we are to stop talking past one another, if communications are to get to the fundamental links between beliefs and actions, new approaches are going to be required.

How about an example? Let us see if we can get at the origins of the author's own attitudes about energy. These are of no particular interest or consequence: I choose them, in fact, partly with the specific intent of exposing what I see to be their total lack of scientific or other absolute claims to special status.

For the sake of this illustration, let us boil the matter down to a few paragraphs. In a nutshell, my personal position might be summarized as follows:

1. Nuclear power and the rapid consumption of fossil fuels buy much that we could well do without; at the same time they exact costs that pose fundamental threats to material sustainability; and they significantly undermine democratic and other desirable strictures of human association.
2. Renewables combined with sharp increases in the efficiency of energy use could provide a materially more sound founda-

tion for human activity while helping us to refocus our attention on things of real importance.

Where, then, do these views come from? I could, of course, marshall the usual collection of facts that would resonate well with my position. But these will not be sufficient to convince those who do not share my views, as they can surely marshall other collections of facts supporting their own opposing views. If we are going to communicate and understand each other, we are going to have to get at more basic issues. My perspective does not arise simply from the facts I may list, but also from very basic personal values and notions of how the world works. Another way to put this is that, although all of the theoretical discussion of earlier chapters is interesting, the questions that established theory raises in my mind appear there because I do not share the beliefs adopted by those who have generated the theory. Again, if we are going to understand each other, we will need to get beyond the usual recitation of competing collections of facts.

Personal values and basic notions of how the world works are not easily communicated. Ultimately it may be that people holding very different views will simply have to get to know each other. Rhetorical forms other than the technical report will be in order. For the sake of the present example, let us again limit the process to a few paragraphs.

Let me focus on the single concept of "strength." Generally speaking, we have every reason to admire strength: in the athlete, in the character of an individual, as the symbol and product of a cultivated discipline, and as a valuable tool in dealing with the vagaries of an uncertain existence. We may not always, on the other hand, admire *impulsive* strength. We do not, in general, admire *impetuous* strength.

For me there is something of impetuous strength in contemporary energy policy and practice. If strength, proper, is analogous to a large bank account, then impulsive strength makes use of the credit card with funds now or soon to be available. Extending the analogy, impetuous strength can be thought of as use of the credit card when, in fact, one cannot afford the purchase: "I want it, so I'll buy it." Impetuous strength is immature, irresponsible, ultimately destructive. The historical trend from "income" energy sources (e.g., wood used roughly at its natural replacement rate), to reliance on "capital" energy sources (e.g., fossil fuels "banked" over millions of years), and finally to a potentially growing dependence on "borrowing" energy sources (e.g., nuclear power whose benefits are received now but whose waste-related and other costs are borrowed against the future) is for me ominous in this context. There is, without doubt, great strength embodied in accumulated human knowledge of nature. But the trend from "income" to "capital," and ultimately to "borrowing," ends with an impetuous application of that strength.

Perhaps we do not live in the proverbial best of all possible worlds. But is the near rejection of the human condition implicit in modern energy policies (and in our often unthinking development and embrace of new technology) appropriate and properly measured? Are we in fact pursuing desirable and carefully considered ends or are we victims of a deep-seated but fundamentally immature insecurity? Have we, as Lewis Mumford suggests, fallen into an impetuous and unthinking rejectionism:

> [It is as if] . . . man has only one all-important mission in life: to conquer nature. By conquering nature the technocrat means, in abstract terms, commanding time and space; and in more concrete terms speeding up every natural process, hastening growth, quickening the pace of transportation, and breaking down communication distances by either mechanical or electronic means. To conquer nature is in effect to remove all natural barriers and human norms and to substitute artificial, fabricated equivalents for natural processes: to replace the immense variety of resources offered by nature by more uniform, constantly available products spewed forth by the machine.
>
> . . . there is only one efficient speed, *faster*; only one attractive destination, *farther away*; only one desirable size, *bigger*, only one rational quantitative goal, *more*.[9]

If we knew how to build fusion energy systems would we have to build them? Can we not pursue knowledge about the world around us without the intention of applying it to alter that world? Is cheap and abundant energy, through fusion or other means, in fact desirable? For what purpose is it so: to support an ever larger population; to "pacify the struggle for existence[10];" to increase the availability of leisure time; to provide opportunity for more meaningful work? For these or other ends, is cheap and abundant energy really the most efficacious means?

I admire and would prefer to emulate the sober strength of my grandparents—strength tempered by firsthand experience of farm life in Kansas, two World Wars, and the Great Depression. Not the weakness of fearful strength. There remains room for humor, for occasional impulsiveness, for life. But it is sober strength. A strength that chooses its way carefully, recognizing that "In nature there are neither rewards nor punishments—there are consequences."[11] A strength that recognizes that not all goals that are achievable are at the same time desirable.

The unwise expend and missapply their strength, and this, in my view, is the essence of our present energy practice. To be truly admirable, strength is likely to hold something in reserve and must be tempered in its application and use by a gentle wisdom.[12]

This is a very brief statement and is, again, intended here only as an illustration of the kind of communication that appears to me to be necessary. The views presented are personal views, of no greater and no less significance than any other person's views, whether the latter are consistent or inconsistent with established energy policy. If this exercise were to be continued, I would probably fill the next 100 pages with childhood memories of my grandfather or offer the reader a copy of C. S. Lewis's essay, "The Abolition of Man" or of Shakespeare's *The Tempest* (see note 12).

The origins of energy perspectives are not easy to know or convey, as the preceding example should demonstrate. They are also ultimately impossible to "defend" in the usual sense and are hardly adequate to "justify" any particular set of public policies. Thinking about them may help us clarify them in our own minds. Efforts to communicate them should do more than any assembly of energy facts to further understanding and our ability to work constructively together. Efforts to communicate and think in this area also amount to more explicit and direct attempts to define "the good life" than the usual, more superficial approach to energy policy debates.[13]

The best communication may be that accomplished through action. The essentially verbal approaches alluded to previously are helpful, but the actions of the home power people, for example, may carry much more weight and detail. Words alone are never fully adequate and seem quickly to spend themselves, whereas actions generate new experience and thus contribute directly to content. Some public support for, or at least tolerance of and interest in, the expression of alternative approaches to energy issues may be appropriate in this light. Certainly denial of the right of expression is profoundly onerous, as the protections of our Bill of Rights recognize; the deep seated commitments evident among home power pioneers in their efforts to develop active alternatives to conventional energy patterns may themselves be indicative of the onerousness of a kind of denial of opportunity for expression through action. Those commitments provide further evidence of the need for improved communication.

Unredeemable Loss

If and as communications are improved, they are likely to reveal fundamental differences. If we no longer insist on the orthodox view, that is, and instead allow all practicable alternatives a more equal voice, it is quite possible that a wide range of fundamentally different approaches to energy issues will arise. All human beings assuredly share certain characteristics and would for this reason have some tendency to cluster in their energy preferences. But if the realm of technical and sociocultural possibility is as broad as has been suggested here, fundamental differences will undoubtedly arise.

Differences of this sort are not new in human affairs. Even a cursory exposure to other cultures or religious traditions today, while it may reveal profound underlying similarities, also reveals fundamental differences. More prolonged exposure may bring a sense of the incommensurability of different traditions and even a sense of divided allegiance as each tradition exerts its independent appeal. Quite apart from energy concerns, we may find valid and profoundly appealing aspects of Japanese, British, American Indian, or other traditions, yet we may be quite unable to imagine a single tradition, or even how to live as an individual, in a way that would combine those traditions into a single existence.

In his introduction to a collection of Isaiah Berlin's essays on the history of ideas, Roger Hausheer refers to a similar situation of incommensurability and has suggested that "in [Berlin's] conception of man and the ends of life there is a powerful element of tragedy: avenues to human realisation may intersect and block one another; things of inestimable intrinsic value and beauty around which an individual or a civilization may seek to build an entire way of life can come into mortal conflict; and the outcome is eradication of one of the protagonists and an absolute unredeemable loss."[14] Whether the differences are historical (as in Berlin's work), contemporary (among nations or cultures), or variously linked to energy issues, there may ultimately be the "necessity for absolute choices." It is impossible for an entire nation to run on coal-fired power plants with all that they imply and, at the same time, to run on roof-mounted photovoltaic power systems with all that they imply.

Where our apprehension of, and response to, energy issues is fundamentally different, "absolute unredeemable losses" will probably be inevitable. The trick, arranged primarily through better communication, perhaps, will be to minimize those losses. The American Indian has not fared well historically. But it is not clear that Americans will have to become like the Japanese in the near future or vice versa. Where commitments to particular energy futures are not mutually exclusive, there may be room for them to coexist. If it is not necessary to insist that everyone rely on fossil fuels or that everyone rely on renewables, perhaps energy futures centered in each can be pursued in parallel.

This will undoubtedly be difficult. Lovins has, for example, suggested that the "hard" and "soft" paths are to a degree mutually exclusive because they inevitably compete for investment resources.[15] Differences and uncertainties will inevitably be threatening to many people, as well.

What is it about relativism that seems to put the fear of god into everyone?

> It is the realization that one's own most cherished point of view may turn out to be just one of many ways of arranging life, important for those brought up in the corresponding tradition, [perhaps] utterly uninteresting and . . . even a hindrance to others. Only few people are content with being able to think and live in a way pleasing to themselves and would not dream of making their tradition an obligation for everyone.[16]

If, through better communication, we are able to gain a deeper understanding of each other's perspectives, there is some prospect for minimizing if not eliminating unredeemable losses as one perspective eclipses others.

Process and Outcome

It is often argued, implicitly, that we cannot afford to experiment or to do what some large or small collection of citizens might wish to do about the energy problem. Certain characteristics of the *outcome* in energy affairs are of critical concern and we cannot afford to indulge in distractions and sidetrips on the way there. Massive political leverage is often generated through arguments of this sort. "We can not do renewables," for example, because the cost penalties would damage the economy, put people out of work, undermine our global competitiveness, and/or lower our standard of living. "We can not continue to build coal-fired power plants," it may be argued, because carbon dioxide emissions already threaten global climates and, by extension, agricultural productivity, coastal populations (through rising sea levels), and the stability of the biosphere. Participatory processes and the distractions of playing around with a variety of possibilities must be set aside in times of crisis, it is implicitly argued, given the critical need to guarantee certain features of the ultimate outcome.

Arguments of this sort often have a disconcertingly technocratic ring to them. "I am the expert," the speaker claims, in effect. "I know" that the economic impact of a commitment to renewables would be disastrous or that global warming is a fact and will lead to biospheric catastrophe. "You may not believe me," and you may be distracted by purported experts with different views, "but these are the facts," and they require certain immediate responses. "We cannot wait" for agreement, or for a population technically sophisticated enough to properly evaluate the issues at hand. "We must act now." Sometimes there is even a genuine, if fundamentally naive, call for "participation" in the implementation of the plan being proferred—a call more like that of a general trying to inspire volunteers for a dangerous mission than that of a true democrat appealing for the full participation of whole citizens.

Occasionally there is evidence of discomfort with this technocratic tendency or with the political leverage exercised along the way. It may be argued, for example, that true participatory approaches to resolving energy problems have a greater likelihood of leading us to good solutions.[17] They have the advantage, after all, of more fully engaging our collective capacity for creative imagination. It could also be argued that the parallel pursuit of several possible energy futures will give us a better chance of arriving at a workable solution than a unified pursuit of one route that might prove to be a dead end. Parallel work on several approaches has produced success in such critical projects as the development of a way to enrich uranium for production of the first atomic bombs during World War II. Where would we be, after all, if we were to adopt the "global warming" theory or the "damaged economy" theory, only to be told well down the road, "Oops! We seem to have made an error."

The arguments presented in this book offer at least weak support for the position that better outcomes are likely to result from a more open, experimental, and (truly) participatory process. If we are able to accept the notion that we do not "know" what to do about energy issues in the traditionally closed sense of scientifically or religiously sanctioned "knowledge," it is easier to imagine the failure of a single approach. Parallel pursuit of a multiplicity of paths would allow us to drop the deadends encountered on particular routes with less than total losses, shifting those who have been on the dead end street across to other interesting possibilities. With no clear cut standard to employ, however, and plenty of uncertainty regarding how any particular standard might be achieved, it is difficult to choose any particular strategy, participatory or technocratic, for addressing energy issues based on the argument that they are more likely to result in the desired outcome.

But there may be a more important question to ask: pushed to the extreme, which is more important, the material *outcome* or the *process* by which choices are made?

In the technocratic response it is assumed, again implicitly, that the outcome concern, the economy or global warming in our examples, is so important that it supercedes issues of process. There is no time to "educate" the ignorant masses, to engage in debate, to experiment with a variety of alternatives. The outcome is critical.

But is it? Clearly, if you are an economist or a climatologist and you look out at the world and (figuratively speaking) see nothing but a collection of lemmings heading straight for the cliff, you confront a difficult dilemma. There is no time for discussion. You either release your aerosol anesthetic or you let them all go off the cliff.

This is an extreme case. And natural and physical scientists may find it a more realistic dilemma than social scientists and those in the humanities because the former are perhaps accustomed to a greater sense of certainty than the latter. But it is, I think, one of the most important questions to consider in deciding how one wishes to address energy issues. It is, in a sense, a very old question: do we go with autocratic rule or with a democratic process? In this form it is a question that our own founding fathers did not answer unambiguously—the selection of both presidents and senators was, for example, not originally entrusted to the people directly.[18] It is a question that goes to the heart of our values, our sense of the role of the individual in society, and our faith or lack of faith in the capacities of our fellow travelers. What do we most want to achieve? Is our focus on people and the development of their capacity for perception and choice or on maintenance of the material necessities of life?

It is not fair, of course, to put this as an either-or question. One would hope that the material necessities of life could generally be maintained without sacrificing opportunities for human development and vice versa. In choosing our strategy, however, it is important to think carefully about where our priorities lie. Is process sometimes more important than outcome? Would it be appropriate in some cases even to abandon oneself to what one sees to be disaster, in the interest of sharing with others the task of defining the problem and choosing the response?

Closing

It may well be easier to abandon oneself to a more open process for dealing with energy issues if one does not need very much and if one is not sure of very much, than if one needs a great deal and knows quite definitely how to get it. Relatively speaking, this author admittedly falls more in the former than in the latter category. Let me nevertheless close with my firmest suspicions.

If we accept the relatively indeterminate image of energy issues presented in this book, if we believe those issues to be of material concern, if we believe that people may be different in their needs and value commitments, then it behooves us to explore more actively the option space available to us. If democracy requires more than one candidate for the future, if, in a democracy, even minority views have a certain claim to breath and to the light of day, then present constraints on the application of collective resources and present barriers to the development of alternatives such as home power lack legitimacy. Energy issues are not reducible to dollars and kilowatt-hours. They are inextricably implicated in how we live our lives. A failure more conscientiously to engage our imaginations in the consideration of energy alternatives for the future is neither democratic nor wise.

Notes

Preface

1. Committee on Nuclear and Alternative Energy Systems (CONAES), "On Thinking About Future Energy Requirements," *Sociopolitical Effects of Energy Use and Policy*, Supporting Paper 5, by Herman E. Daly (Washington, D.C.: National Academy of Sciences, 1979), pp. 231–241.

2. Johan W. Schot, "Constructive Technology Assessment and Technology Dynamics: The Case of Clean Technologies," *Science, Technology, and Human Values* 17 (Winter 1992): 36–56.

Chapter 1. Raising the Question

1. Bruno Bettelheim, *The Informed Heart: Autonomy in a Mass Age* (New York: Avon Books, 1960).

2. Ibid., pp. 81–83.

3. Ibid., pp. 154–155.

4. Paul C. Stern and Elliot Aronson, eds., *Energy Use: The Human Dimension* (New York: W. H. Freeman and Company, 1984).

5. Ibid., p. 23.

6. Langdon Winner has suggested that, although "values" do play an important role in the bureacratic and technocratic vocabulary,

they are defined "in that lexicon [as] something like 'the residual concerns one needs to ponder after the real, practical business of society has been taken care of.'" See Langdon Winner, *The Whale and the Reactor* (Chicago: University of Chicago Press, 1986), p. 160.

7. A similar "invisibility" in the much broader context of our way of taking up with the world through technology has been widely noted. See, for example, Langdon Winner, *Autonomous Technology* (Cambridge, Mass.: MIT Press, 1977), p. 6; Albert Borgmann, *Technology and the Character of Contemporary Life* (Chicago: University of Chicago Press, 1984), p. 35; or Martin Heidegger, *The Question Concerning Technology*, trans. W. Lovitt (New York: Harper and Row, 1977).

8. Stern and Aronson, p. vii (emphasis added).

9. Noam Chomsky has argued, on the other hand, that the burden of proof should lie with those who believe that controls are necessary and that a more open "society" is impossible—i.e., that we should not assume ab initio that a society cannot exist without (or with much less in the way of) restriction of thought and action. (Personal communication during Chomsky's visit at Michigan Technological University, Houghton, Michigan, March 27, 1990.)

10. Winner, *The Whale*, p. 27. In this quotation, Winner is referring specifically to the development of a mechanical tomato harvester well-suited only for large tomato growing operations.

11. A number of recent authors appear to have adopted analogous notions of effective conspiracy. Compare, for example, Winner, *The Whale*; John Galbraith, *The New Industrial State* (Boston: Houghton Mifflin Company, 1967); David Noble, *America by Design: Science, Technology, and the Rise of Corporate Capitalism* (New York: Alfred A. Knopf, 1977); Harry Braverman, *Labor and Monopoly Capital: The Degradation of Work in the Twentieth Century* (New York: Monthly Review Press, 1974).

12. Compare Hughes's notion of "momentum" in Thomas P. Hughes, "The Evolution of Large Technological Systems" in Wiebe E. Bijker, Thomas P. Hughes, and Trevor Pinch, eds., *The Social Construction of Technological Systems* (Cambridge, Mass.: MIT Press,

1989) and Winner's notion of technological "somnambulance" in *The Whale.*

13. I am thinking here of democracy in the sense of Feyerabend's "free society." See Paul Feyerabend, *Science in a Free Society* (London: Verso Editions/NLB, 1978).

14. Elting E. Morison, *From Know-How to Nowhere: The Development of American Technology* (New York: Basic Books, 1974) pp. 178–179.

15. For an excellent presentation of links between energy preferences and different ways of looking at things, see Energy Research and Development Administration, *Solar Energy in America's Future,* Chapter 5: "Broader Issues," by Willis Harman et al. (prepared by Stanford Research Institute International for the Energy Research and Development Administration, March 1977, DSE-115/1).

16. See, for example, the compilation of renewable energy projects in Kathleen Courrier et al., eds., *Shining Examples* (Washington, D.C.: Center for Renewable Resources, 1980) or Nancy Rader, "The Power of the States: A Fifty-State Survey of Renewable Energy," *Public Citizen,* (June 1990).

Chapter 2. So Much Fuss over Energy

1. See, for example, Committee on Nuclear and Alternative Energy Systems (National Academy of Sciences), *Energy in Transition: 1985–2010* (San Francisco: W. H. Freeman and Company, 1979).

2. It may be worth emphasizing that these energy inputs must be in the form of actual calories, BTUs, and kilowatt-hours, not simply "conserved energy." It has been fashionable to regard conservation or efficiency improvements as a source of supply. See, for example, Robert Stobaugh and Daniel Yergin, *Energy Future* (New York: Vintage Books, 1983), p. 173: "There is a *source* of energy that produces no radioactive waste, nothing in the way of petrodollars, and very little pollution. Moreover, the *source* can provide the energy that conventional sources may not be able to furnish. The *source* might be called energy efficiency, for Americans like to think of themselves as

an efficient people. But the energy *source* is generally known by the more prosaic term *conservation*. To be semantically accurate, the *source* should be called conservation energy, to remind us of the reality—that conservation is no less an energy alternative than oil, gas, coal, or nuclear." (Emphasis added to each use of the word *source*). Such an approach may have short range benefits in terms of encouraging a shift of new investment from traditional coal, oil, gas, and other supply systems more toward investment in efficiency improvements. It has obvious limits in the sense that one cannot ultimately eat efficiency improvements or use them directly to warm a home. In the end one does need an actual *material* "source" of energy.

3. Carol E. Steinhart and John S. Steinhart, "The Energy We Eat," in *Energy: Sources, Use, and Role in Human Affairs* (North Scituate, Mass.: Duxbury Press, 1974), pp. 65–88.

4. Energy Information Administration, *Monthly Energy Review* (Washington, D.C.: U.S. Department of Energy, January 1994), DOE/EIA-0035(94/01), p. 113.

5. Ibid., p. 15.

6. For a brief and highly readable account of major events in the struggle for control of world oil markets from the early 1900s to the mid 1980s, see Stobaugh and Yergin, *Energy Future*. For a more exhaustive treatment, see Daniel Yergin, *The Prize* (New York: Simon and Schuster, 1991).

7. See, for example, Federal Energy Administration, *Project Independence Report* (Washington, D.C.: U.S. Government Printing Office, GPO Stock No. 4118-00029, 1974); Energy Research and Development Administration, *Creating Energy Choices for the Future* (Washington, D.C., U.S. Energy Research and Development Administration, ERDA-48, June 30, 1975); Carroll L. Wilson, project director, *ENERGY: Global Prospects 1985–2000*, Report of the Workshop on Alernative Energy Strategies Sponsored by the Massachusetts Institute of Technology (New York: McGraw-Hill Book Company, 1977); Executive Office of the President (Jimmy Carter), "The National Energy Plan" (Washington, D.C.: U.S. Government Printing Office, GPO Stock No. 040-000-00380-1, April 1977); Sam H. Schurr et al., *Energy in America's Future: The Choices Before Us*

(Baltimore: Johns Hopkins University Press, 1979); Committee on Nuclear and Alternative Energy Systems, *Energy in Transition*; Solar Energy Research Institute, *A New Prosperity: Building a Sustainable Energy Future* (Andover, Mass.: Brick House Publishing, 1981); Paul C. Stern and Elliot Aronson, eds., *Energy Use: The Human Dimension*, National Research Council report (New York: W. H. Freeman and Company, 1984); U.S. Department of Energy (Reagan administration), "The National Energy Policy Plan" (Washington, D.C.: U.S. Department of Energy, DOE/S-0040, 1985). One private and three other official studies are also worth noting for the departures they include from the mainstream thought of the time: Ford Foundation, *A Time to Choose: America's Energy Future*, Final Report by the Energy Policy Project of the Ford Foundation (Cambridge, Mass.: Ballinger Publishing Co., 1974); Hans H. Landsberg et al., *Energy and the Social Sciences: An Examination of Research Needs*, A Resources for the Future Study Supported by the National Science Foundation, Contract GI-37032 (Washington, D.C.: Resources for the Future, Inc., July 1974); Energy Research and Development Administration, Willis Harman et al,. *Solar Energy in America's Future*, (Prepared by Stanford Research Institute International for the Energy Research and Development Administration, March 1977, DSE-115/1), Chapter 5: "Broader Issues"; and Paul Craig et al., "Distributed Energy Systems in California's Future: Interim Report" (Washington, D.C.: U.S. Department of Energy, HCP/P7405-03, 1978). The last of these appeared only in the form of an interim report since the research program from which it emerged was terminated midstream.

8. See, for example, Federal Energy Administration, *Project Independence Report.*, from the Nixon administration.

9. For a good general survey of the status and prospects for energy efficiency improvements, see Stobaugh and Yergin, *Energy Future*; Solar Energy Research Institute, *A New Prosperity*; or *Energy for Planet Earth*, Readings from *Scientific American* (New York: W. H. Freeman and Company, 1991). More specific studies of energy savings potential may also be of interest, including work on long-range prospects like Amory Lovins, John W. Barnett, and L. Hunter Lovins, "Supercar: The Coming Light-Vehicle Revolution" (Old Snowmass,

Colo.: Rocky Mountain Institute, March 1993), covering efficiency limits for the automobile.

10. According to Stobaugh and Yergin, ibid., p. 173, for example, "If the United States were to make a serious commitment to conservation, it might well consume perhaps 30 percent less energy than it now does, and still enjoy the same or an even higher standard of living. That saving would not hinge on a major technological breakthrough, and it would require only modest adjustments in the way people live. Moreover, the cost of conservation energy is very competitive with other energy sources. The possible energy savings would be the equivalent of the elimination of all imported oil—and then some."

11. For a number of years, Amory Lovins gave talks around the country that included a presentation of how official projections of future energy use had shifted downward. What was considered "beyond the pale" in the early 1970s, in the sense of being an impossibly low estimate of future energy use, came to be roughly in the range of official estimates as little as five or ten years later. This presentation has been captured in a videotape, "Lovins on the Soft Path: An Energy Future with a Future" (1982), available from Bullfrog Films or through the Rocky Mountain Institute, Old Snowmass, Colorado. At least as late as the 1979 CONAES study (Committee on Nuclear and Alternative Energy Systems, *Energy in Transition*, p.8) scenarios for future energy use included extrapolations at the 3.5 percent per year growth rates observed for the period 1950 to 1973. More recent official projections include energy growth rates at half the rate of growth of GNP, with energy growth declining over the next forty years from 1.6 percent in the period 1990–2000 down to 0.9 percent in the period 2020–2030 (U.S. Department of Energy, *National Energy Strategy*, (Washington, D.C.: U.S. Department of Energy, DOE/S-0082P, February 1991).

12. Committee on Nuclear and Alternative Energy Systems, *Energy in Transition*, p.5.

13. C. D. Masters, D. H. Root, and E. D. Attanasi, "World Oil and Gas Resources—Future Production Realities" *Annual Review of Energy* 15 (1990): 23–51.

14. See, for example, Schurr et al., *Energy in America's Future.*

15. Gordon MacDonald, "The Future of Methane as an Energy Resource," *Annual Review of Energy* 15 (1990): 53–83.

16. Schurr et al., *Energy in America's Future*, p. 226; Committee on Nuclear and Alternative Energy Systems, *Energy in Transition*, pp. 46-47.

17. *Energy for Planet Earth*, p. 111.

18. Stephen Schneider, "The Changing Climate," *Managing Planet Earth* (New York: W. H. Freeman and Company, 1990).

19. See, for example, Nancy Kubasek and Gary Silverman, *Environmental Law* (Englewood Cliffs, N.J.: Prentice-Hall, 1994).

20. For brief outlines of many of these problems, see Walter Rosenbaum, *Environmental Politics and Policy*, 2d ed. (Washington, D.C.: CQ Press, 1991).

21. John Holdren, "Energy in Transition," *Energy for Planet Earth*, p. 123. See also Lester R. Brown and Sandra Postel, "Thresholds of Change," *State of the World 1987* (New York: W. W. Norton and Company, 1987), pp. 3–19.

22. For discussion of a number of energy-related equity issues see Stern and Aronson, *Energy Use*.

23. Lewis Mumford, "Authoritarian and Democratic Technics," *Technology and Culture* 5 (1964): 1–8.

24. Ratio based on data in Paul Ehrlich, Ann Ehrlich, and John Holdren, *Ecoscience* (San Francisco: W. H. Freeman and Company, 1977).

25. Lester Brown and Sandra Postel ("Thresholds of Change") have made arguments along roughly this line, also citing the historical decline of Mesopotamian and Mayan civilizations in the context of related energy and environmental concerns. One could also think of the Great Depression of 1929 as an almost pure illustration of the impact of disruptions in complex systems of specialized interdependence; in this case, in fact, there were not even any substantial resource-related constraints as the recovery of the Roosevelt era amply demonstrated.

26. Arguments examined in Chapter 4 could be taken to suggest even that the recent collapse of the Soviet Union and the global economic slowdown of the 1990s are both ultimately traceable to the

declining availability of high yield energy resources. Such an argument would suggest that we are already seeing the sorts of negative disruptions alluded to in the text. The collapse of organization in the nation of Somalia and the ensuing mass starvation and international intervention of 1992–1993 may also offer an illustration of the threats inherent simply in the destabilization of production systems.

27. See, for example, Brown and Postel, "Thresholds of Change."

28. Langdon Winner, *The Whale and the Reactor: A Search for Limits in an Age of High Technology* (Chicago: University of Chicago Press, 1986), p. 12.

29. As a matter of speculation, it is interesting to wonder whether the recent popularity of such things as backpacking might stem to some degree from a sense that such a perspective is needed.

30. Winner, *The Whale*, pp. 22–23.

31. Ibid., p. 47.

32. Ibid., pp. 47–48.

33. Ibid., pp. 3–18.

34. David Strong, "The Technological Subversion of Environmental Ethics," *Research in Philosophy and Technology: Technology and the Environment*, ed. Frederick Ferre (Greenwich, Conn.: JAI Press, 1992), pp. 33–66.

35. Thomas P. Hughes, "The Evolution of Large Technological Systems," in *The Social Construction of Technological Systems*, ed. Wiebe E. Bijker, Thomas P. Hughes, and Trevor Pinch (Cambridge, Mass.: MIT Press, 1989), pp. 51–82.

36. For a more detailed discussion of agriculture along this line, see Wendell Berry, *The Unsettling of America* (New York: Avon Books, 1977).

37. Amory Lovins, "Energy Strategy: The Road Not Taken?" *Foreign Affairs* 55, no. 1 (1976): 91–92.

38. Ibid., p. 94.

39. Winner, *The Whale*, pp. 19–39.

40. Russell W. Ayres, "Policing Plutonium: The Civil Liberties Fallout," *Harvard Civil Rights–Civil Liberties Law Review* 10, no. 2 (Spring 1975): 369–443.

41. For example, we might well find a world of 12 billion people a less than desirable world in which to live.

42. E. F. Schumacher, "The Age of Plenty: A Christian View," *Economics, Ecology, Ethics: Essays Toward A Steady-State Economy*, ed. Herman E. Daly (San Francisco: W. H. Freeman and Company, 1980), p. 136.

43. Alvin Weinberg, "Can Technology Replace Social Engineering?" in *Technology and the Future*, ed. Albert H. Teich (New York: St. Martin's Press, 1990), pp. 29–38.

44. One could carry analysis of Weinberg's classic article further. It is not clear, for example, that social engineering really is less tractable. More important, Weinberg fails to recognize that the "technological fix," itself, can be seen as one of the most formidable forms of social engineering (at least retrospectively) that one might imagine.

45. See, for example, "Energy Conservation" (entire volume), *Journal of Social Issues* 37, no. 2 (Spring 1981).

46. Mumford, "Authoritarian and Democratic Technics."

47. Winner, *The Whale*, pp. 10 and 169.

48. See, for example, Mumford, "Authoritarian and Democratic Technics," or C. S. Lewis, "The Abolition of Man," in *Economics, Ecology, Ethics*, pp.177–187.

Chapter 3. Surely the Experts Have Thought the Matter Out

1 Possible exceptions include elements of the final report of the Energy Policy Project of the Ford Foundation, *A Time to Choose: America's Energy Future* (Cambridge, Mass.: Ballinger Publishing Co., 1974) and Energy Research and Development Administration, *Solar Energy in America's Future*, Willis Harman et al., (Prepared by Stanford Research Institute International for the Energy Research and Development Administration, DSE-115/1, March 1977), Chapter 5: "Broader Issues."

2. See, for example, Federal Energy Administration, *Project Independence Report* (Washington, D.C.: U.S. Government Printing Office, GPO Stock No. 4118-00029, November 1974); Sam H. Schurr et al., *Energy in America's Future: The Choices Before Us* (Baltimore:

Johns Hopkins University Press, 1979); Committee on Nuclear and Alternative Energy Systems (National Academy of Sciences), *Energy in Transition: 1985–2010* (San Francisco: W. H. Freeman and Company, 1979); United States Department of Energy (Reagan administration), *The National Energy Policy Plan*, (Washington, D.C.: U.S. Department of Energy, DOE/S-0040, 1985); United States Department of Energy, *National Energy Strategy: Powerful Ideas for America* (Washington, D.C.: U.S. Government Printing Office, February 1991).

3. Palmer C. Putnam (consultant to the Atomic Energy Commission), *Energy in the Future* (Toronto: D. Van Nostrand Company, 1953).

4. Lewis Mumford, "Authoritarian and Democratic Technics," *Technology and Culture* 5 (1964): 1–8.

5. Committee on Nuclear and Alternative Energy Systems, *Energy in Transition.*

6. Ibid., p. 66.

7. Laura Nader and Stephen Beckerman, "Energy as It Relates to the Quality and Style of Life," *Annual Review of Energy* 3 (1978): 1–28.

8. National Research Council Board on Atmospheric Sciences and Climate, *Changing Climate: Report of the Carbon Dioxide Assessment Committee* (Washington, D.C.: National Academy Press, 1983), p. 51.

9. Ibid., p. 481.

10. T. P. Hughes might use the term *reverse salients* here. See Thomas P. Hughes, "The Evolution of Large Technological Systems," in *The Social Construction of Technological Systems*, ed. Wiebe E. Bijker, Thomas P. Hughes, and Trevor Pinch (Cambridge, Mass.: MIT Press, 1989).

11. Committee on Nuclear and Alternative Energy Systems, *Energy in Transition*, p. 43.

12. Ibid., p. 619. Emphasis added.

13. Herman E. Daly, "On Thinking About Future Energy Requirements," in *Sociopolitical Effects of Energy Use and Policy*, Committee on Nuclear and Alternative Energy Systems (CONAES)

Supporting Paper 5 (Washington, D.C.: National Academy of Sciences, 1979), p. 237.

14. Ibid., p. 236.

15. For a concise quantitative and qualitative introduction to the practice of discounting, see Donald M. McAllister, *Evaluation in Environmental Planning: Assessing Environmental, Social, Economic, and Political Trade-offs* (Cambridge, Mass.: MIT Press, 1986).

16. Schurr et al., *Energy in America's Future*, p. 355.

17. Richard B. Norgaard, "Economic Indicators of Resource Scarcity: A Critical Essay," *Journal of Environmental Economics and Management* 19 (1990): 19–25.

18. Ibid., p. 24.

19. Harold Hotelling, "The Economics of Exhaustible Resources," *Journal of Political Economy* 39, no. 2 (April 1931): 137–175; Anthony C. Fisher, *Resource and Environmental Economics* (Cambridge: Cambridge University Press, 1981).

20. Hotelling, ibid.

21. Fisher, *Resource and Environmental Economics*; P. S. Dasgupta and G. M. Heal, *Economic Theory and Exhaustible Resources* (Cambridge: Cambridge University Press, 1979); William Nordhaus, "The Allocation of Energy Resources," *Brookings Papers on Economic Activity* 3 (1973): 529–570.

22. The assumption that there are no serious externalities is made at least implicitly in all of the references of notes 19 and 21. A few analysts have called economists to task for the flimsy handling of this point; see, for example, William J. Baumol, "External Economies and Second-Order Optimality Conditions," *American Economic Review* 54, no. 4 (June 1964): 358–372.

23. Committee on Nuclear and Alternative Energy Systems, *Energy in Transition*, p. vi.

24. Elizabeth Colson, "Tranquility for the Decision Maker," in *Cultural Illness and Health*, ed. Laura Nader and Thomas W. Maretzki (Washington, D.C.: American Anthropological Association, 1973).

25. Daly, "On Thinking About Future Energy Requirements."

26. Ibid., p. 232.

27. Ibid., p. 232.

Chapter 4. With Troubles Enough, Experts Differ

1. See, for example, Sam H. Schurr et al., *Energy in America's Future: The Choices Before Us* (Baltimore: Johns Hopkins University Press, 1979).

2. See, for example, Daniel Yergin, "Conservation: The Key Energy Source," in *Energy Future*, ed. Robert Stobaugh and Daniel Yergin (New York: Random House, 1983), pp. 179–182; Energy Information Administration, *Monthly Energy Review* (Washington D.C.: U.S. Department of Energy, March 1989), pp. 13, 14, 26.

3. Arguments with this general thrust go back to those of F. Soddy, *Wealth, Virtual Wealth and Debt: The Solution of the Economic Paradox* (New York: G. P. Dutton, 1933) and Fred Cottrell, *Energy and Society: The Relation Between Energy, Social Change, and Economic Development* (Westport, Conn.: Greenwood Press, 1955). They have been advanced more recently by Nicholas Georgescu-Roegen, *Energy and Economic Myths* (New York: Pergamon Press, 1976) and Richard N. Adams, *Energy and Structure* (Austin: University of Texas Press, 1975). Such arguments are also supported by others discussed and cited later in the text.

4. John Gever, Robert Kaufmann, David Skole, and Charles Vorosmarty, *Beyond Oil: The Threat to Food and Fuel in the Coming Decades* (Cambridge, Mass.: Harper and Row, Publishers, 1986), p. 41.

5. Cutler Cleveland, Robert Costanza, Charles A. S. Hall, and Robert Kaufmann, "Energy in the U.S. Economy: A Biophysical Perspective," *Science* 225 (August 1984): 890–897. Citations for internal quotes are R. M. Solow, *American Economist* 2, no. 5 (1978) and W. Nordhaus and J. Tobin in *The Measurement of Economic and Social Performance*, ed. M. Moss (New York, Columbia University Press, 1973).

6. Robert Costanza, "Embodied Energy and Economic Valuation," *Science* 210 (December 12, 1980): 1219–1224.

7. Ibid., p. 1223.

8. Ibid.

9. Cleveland et al., "Energy in the U.S. Economy," p. 893.

10. Cottrell, *Energy and Society*.

11. Depending upon the precise dynamics of a transition to renewables, it can even be argued that the social disruptions associated with increases in the cost of energy relative to people's time could best be minimized by an early transition to renewables. The more depleted conventional energy sources are at the time a transition is made, the more precious those conventional sources will be as some portion of them is diverted to the construction of new (renewable) energy supply systems.

12. Cleveland et al., "Energy in the U.S. Economy," p. 896.

13. Gever et al., *Beyond Oil*, p. 219.

14. Cleveland et al., "Energy in the U.S. Economy," p. 896.

15. Herman E. Daly, "Introduction to the Steady-State Economy," in *Economics, Ecology, Ethics: Essays Toward a Steady-State Economy*, ed. Herman E. Daly (San Francisco: W. H. Freeman and Company, 1980), p. 19.

16. Ibid., p. 22.

17. E. F. Schumacher, "The Age of Plenty: A Christian View" in *Economics, Ecology, Ethics*, p. 131.

18. Herman Daly and John B. Cobb, *For the Common Good: Redirecting the Economy Toward Community, the Environment, and a Sustainable Future* (Boston: Beacon Press, 1989).

19. Daly, "Introduction,", p. 12.

20. Gerald Garvey. "Research on Energy Policy: Processes and Institutions," in *Energy and the Social Sciences: An Examination of Research Needs*, ed. Hans H. Landsberg et al. (A Resources for the Future Study, Working Paper EN-3, Supported by NSF Grant GI-37032) (Washington, D.C.: Resources for the Future, 1974), pp. 539–580.

21. M. I. Finley, *The Ancient Economy* (Berkeley: University of California Press, 1973), p. 21.

22. Ibid., p. 22.

23. Ibid., p. 40

24. Ibid., p. 84.

25. Marshall Sahlins, *Stone Age Economics* (New York: Aldine Publishing Company, 1972).

26. Compare, for example, the "great transformation" of European civilization from essentially domestic modes of production to industrial economies in Karl Polanyi, *The Great Transformation* (Boston: Beacon Press, 1944).

27. Sahlins, *Stone Age Economics*, p. 39.

28. Committee on Nuclear and Alternative Energy Systems (National Academy of Sciences), *Energy in Transition: 1985–2010* (San Francisco: W. H. Freeman and Company, 1979).

29. Laura Nader, "Barriers to Thinking New About Energy," *Physics Today* (February 1981): 99.

30. Martha W. Gilliland, "Energy Analysis and Public Policy," *Science* 189 (September 1975): 1051–1056; Max R. Langham, W. W. McPherson, Henry M. Peskin, Robert F. Mueller, David E. Reichle, and Martha W. Gilliland, "Energy Analysis," *Science* 192 (April 2, 1976): 8–12; David A. Huettner, "Net Energy Analysis: An Economic Assessment," *Science* 192 (April 9, 1976): 101–104. For a brief time, "net energy analysis" of new energy sources was federally mandated under the Non-Nulcear Energy Research and Development Act of 1974. (See Martha W. Gilliland, "Energy Analysis and Public Policy," p. 1055.)

31. These questions have received growing attention in general and with application to other specific contexts beginning roughly with the work of Thomas S. Kuhn, *The Structure of Scientific Revolutions* (Chicago: The University of Chicago Press, 1962) and culminating more recently with work on the "social construction of science" by Bruno Latour, *Science in Action* (Cambridge, Mass.: Harvard University Press, 1987) and others. A sampling of works in this expanding area of literature might include: David Bloor, *Knowledge and Social Imagery* (Chicago: The University of Chicago Press, 1976); Marcel C. LaFollette, *Making Science Our Own: Public Images of Science 1910-1955* (Chicago: The University of Chicago Press, 1990); and David Dickson, *The New Politics of Science* (Chicago: The University of Chicago Press, 1984).

Chapter 5. The Shaping of Responses

1. Harold Lasswell, *Politics: Who Gets What, When and How* (New York: McGraw-Hill, 1936).

2. See the works of Michel Foucault. For a very brief statement, see Alessandro Fontana and Pasquale Pasquino (interviewers), "Truth and Power," *Power/Knowledge: Selected Interviews and Other Writings 1972–1977*, ed. Colin Gordon (New York: Pantheon Books, 1980), pp. 109–133. See especially pages 131–133.

3. What is actually offered here is, in a sense, an interpretation of technological change, or the lack thereof, in the energy sector. This interpretation owes something to the models of technology development proposed by historians, philosophers, and sociologists of technology, such as T. P. Hughes, *Networks of Power: Electrification in Western Society, 1880–1930* (Baltimore: Johns Hopkins University Press, 1983); Albert Borgmann, *Technology and the Character of Contemporary Life* (Chicago: University of Chicago Press, 1984); Bruno Latour, *Science in Action* (Cambridge, Mass.: Harvard University Press, 1987); and Wiebe E. Bijker, Thomas P. Hughes, and Trevor Pinch, eds., *The Social Construction of Technological Systems* (Cambridge, Mass.: MIT Press, 1987). It is also derived in a broad sense from the works of Lewis Mumford, especially *Technics and Human Development* and *The Pentagon of Power (The Myth of the Machine*, vols. 1 and 2) (New York: Harcourt Brace Jovanovich, 1966 and 1964, respectively); and Langdon Winner, *Autonomous Technology: Technics-out-of-Control as a Theme in Political Thought* (Cambridge, Mass.: MIT Press, 1977) and *The Whale and the Reactor* (Chicago: University of Chicago Press, 1986). It is also informed both by my own reading and observation of developments since the early 1970s and by conversations with other observers, Professors Mark Christensen and Laura Nader and the other faculty and graduate students of the Energy and Resources Group at the University of California at Berkeley.

4. Winner, *The Whale*, p. 10. A number of ways of characterizing the practical situation could be developed here with perhaps the same

accuracy and appeal as Winner's notion of somnambulance. Another with particular appeal, I believe, is the notion that the institutional momentum and popular nonparticipation we observe in connection with energy decision making are reflections of a "hegemonic world-view" that blocks action alternatives from our consciousness. The theoretical work of Antonio Gramsci may be useful in developing an interpretation along these lines as sketched later in this chapter or as outlined in Jesse Tatum, "Energy and Society: Beyond the Bounds of Conventional Analysis," doctoral dissertation, University of California, Berkeley (Ann Arbor, Mich.: University Microfilms International, 1988). See also W. Adamson, "Hegemony, Historical Bloc, and Italian History," *Hegemony and Revolution, Antonio Gramsci's Political and Cultural Theory* (Berkeley: University of California Press, 1980); and Jurgen Habermas, "Technology and Science as 'Ideology,'" trans. Jeremy J. Shapiro, *Toward a Rational Society* (Boston: Beacon Press, 1970).

5. See "Removing Carbon Dioxide From Power Plant Emissions," the newsletter of the MIT Energy Laboratory April–September 1989).

6. Ibid.

7. I am using the term "momentum" in much the way it is used by T. P. Hughes. See, for example, his chapter, "The Evolution of Large Technological Systems," in Bijker et al., *Social Construction*, pp. 51–82.

8. A few exceptional privately funded studies did not fully adopt the standard assumptions, notably the Ford Foundation study completed soon after the 1973 OPEC oil embargo: Ford Foundation, *A Time to Choose: America's Energy Future*, Final Report by the Energy Policy Project of the Ford Foundation (Cambridge, Mass.: Ballinger Publishing Co., 1974).

9. See, for example, Federal Energy Admnistration, *Project Independence Report* (Washington D.C.: U. S. Government Printing Office, GPO Stock No. 4118-00029, November 1974); Carroll L. Wilson, project director, *Energy: Global Prospects 1985–2000*, Report of the Workshop on Alternative Energy Strategies Sponsored by the Massachusetts Institute of Technology (New York: McGraw-Hill Book

Company, 1977); Sam H. Schurr et al., *Energy in America's Future: The Choices Before Us* (Balitimore: Johns Hopkins University Press, 1979); Committee on Nuclear and Alternative Energy Systems (National Academy of Sciences), *Energy in Transition: 1985–2010* (San Francisco: W. H. Freeman and Company, 1979). See also the summary of projections provided by Amory Lovins in "Lovins on the Soft Path: An Energy Future with a Future, A Guide to the Film for Students and Teachers," transcript of portions of public address by Lovins (Old Snowmass, Colo.: Rocky Mountain Institute, 1985), pp. 10–11.

10. Amory Lovins, "Energy Strategy: The Road Not Taken?" *Foreign Affairs* 55. no. 1 (October 1976): 65–96.

11. Some foreshadowing of Lovins's work can be seen in, for example, Ford Foundation, *A Time to Choose*. Lovins, however, coined the term and added substantially to the definition of a "soft path" option, and first brought this alternative to broadened popular attention.

12. Lovins, "Energy Strategy," p. 67.

13. Ibid., p. 71.

14. Elements of this history are covered in Walter A. Rosenbaum, *Energy Politics and Public Policy*, 2d ed. (Washington, D.C.: Congressional Quarterly Inc., 1987), e.g., pages 21 and 90, although he does not explicitly focus on Lovins's role.

15. Committee on Nuclear and Alternative Energy Systems, *Energy in Transition*.

16. Laura Nader, "Barriers to Thinking New About Energy," *Physics Today* (February 1981): 99.

17. Ibid., p. 99.

18. Laura Nader, personal communication, Berkeley, California. Although this rationalization is understandable, it may not be excusable.

The discussion at this point bears some resemblance to the "seamless web" model offered by Thomas P. Hughes, *Networks of Power*, and recently ascribed more broadly to "systems," "social constructivist," and "actor network" approaches current in the field of science, technology and society studies. (See also Johan W. Schot, "Constructive Technology Assessment and Technology Dynamics:

The Case of Clean Technologies," *Science, Technology, and Human Values* 17, no. 1 (Winter 1992): 36–56, especially p. 38.)

19. Rosenbaum, *Energy Politics*, p. 90.

20. Energy Information Administration, *Monthly Energy Review* (Washington, D.C.: Department of Energy, November 1991).

21. See, for example, United States Department of Energy, *National Energy Strategy: Powerful Ideas for America* (Washington, D.C.: February 1991), and the subsequent, only slightly moderated, policies of the Clinton administration.

22. The notion of "scarcity" as the organizing principle of the modern industrial world is worthy of study in itself. Especially interesting synthetic work in this area has been done recently by Pieter Tijmes. See his forthcoming essay, "Why Nature is Perceived as a Meagre Provider: A Social Constructivist Approach," *Concordia*, 26 (1994): 97–113 and his related work, Reginald Luyf and Pieter Tijmes, "Modern Immaterialism," *Research in Philosophy and Technology: Technology and the Environment*, 12 (1992): 271–287. Tijmes serves on the faculty of philosophy and social sciences, University of Twente, Enschede, The Netherlands.

23. In Steven Lukes's terms, I am arguing the existence of "counterfactuals" to the hypothesis that people actually prefer existing energy patterns and policies. See Steven Lukes, *Power: A Radical View* (London: Macmillan Press Ltd., 1974).

24. B. C. Farhar, P. Weis, C. T. Unseld, and B. Burns, "Public Opinion About Energy: A Literature Review" SERI/TR 53-155 (Golden, Colo.: Solar Energy Research Institute, 1979).

25. Council on Environmental Quality, "Solar Energy: Progress and Promise" (Washington, D.C.: U.S. Government Printing Office, GPO Stock No. 041-011-00036-0, April 1978).

26. Matthew L. Wald, "By 2005, Nuclear Unit Sees 50–50 Chance of Meltdown," *New York Times* (April 17, 1985), p. A16.

27. See Jesse Tatum and T. K. Bradshaw, "Energy Production by Local Governments: An Expanding Role," *Annual Review of Energy* 11 (1986): 471–512.

28. Again, this amounts to an argument for the existence of "counterfactuals" in Steven Lukes's terms. See Lukes, *Power*.

29. See Adamson, "Hegemony, Historical Bloc, and Italian History."

30. "Technocratic concept of progress" is a term used by Leo Marx in "Does Improved Technology Mean Progress?" reprinted from *Technology Review* (January 1987) in *Technology and the Future*, Albert Teich, ed. (New York: St. Martin's Press, 1990), pp. 3–14. The term "promise of technology" is used by Albert Borgmann in *Technology and the Character of Contemporary Life* (Chicago: University of Chicago Press, 1984). The discussion that follows relies primarily on Borgmann's formulation of the "promise of technology."

31. Borgmann, ibid., p. 76.

32. Ibid, p. 38.

33. Ibid, p. 38.

34. Ibid, p. 42.

35. Steven Lukes, *Power*.

36. Ibid, p. 51.

Chapter 6. Exploring the Option Space

1. Smelser's term "value-oriented movement," seems an accurate label for the home power movement. See Neil J. Smelser, *Theory of Collective Behavior* (New York: Macmillan Publishing Co., 1962).

2. Amory Lovins, *Soft Energy Paths: Toward a Durable Peace* (Cambridge, Mass.: Ballinger Publishing Company, 1977), pp. 22–23.

3. Russell Ayres, "Policing Plutonium: The Civil Liberties Fallout," *Harvard Civil Rights–Civil Liberties Law Review* 10, no. 2 (Spring 1975): 369–443.

4. Kenneth E. Boulding, "General Comments" in Committee on Nuclear and Alternative Energy Systems (National Academy of Sciences), *Energy in Transition 1985–2010* (San Francisco: W. H. Freeman and Company, 1979), pp. 613–618.

5. Much of the remainder of this chapter is taken from two previously published articles: Jesse Tatum, "Technology and Values: Getting Beyond the 'Device Paradigm' Impasse," *Science, Technology, and Human Values* 19, no. 1 (Winter 1994): 70–87; and Jesse Tatum, "The Home Power Movement and the Assumptions of Energy Policy

Analysis," *Energy—The International Journal* 17, no. 2 (February 1992): 99–108.

6. The 1989 estimate, along with several others and their sources, are outlined in Jesse Tatum, "The Home Power Movement." The more recent estimate of 100,000 systems appears to originate with John Schaeffer, president of the Real Goods Trading Corporation of Ukiah, California. This number has also appeared with or without attribution in John Greenwald, "Here Comes the Sun," *Time* (Oct. 18, 1993): 84–85; in Hannah Holmes, "Unplugged," *Sierra* 78, no. 5 (September–October 1993): 23–24; and in Michael Potts, *The Independent Home* (Post Mills, Vt.: Chelsea Green Publishing Company, 1993).

7. Even the recognition that begins to be evident in 1993 fails adequately to capture the nature and uniqueness of the movement. An article in *Time* (Greenwald, "Here Comes the Sun") for example, begins with mention of the expanding home power supplier, the Real Goods Trading Corporation, but then falls back into traditional notions of declining costs approaching competitive levels. The article ends by focusing on utility mega projects. Such approaches may well prevail over the home power approach in the end. The point remains that at a popular and policy level the activities and nature of the home power movement continue to go largely unrecognized.

8. Federal programs have, nevertheless focused almost exclusively on very large PV arrays and appear to proceed from the unquestioned assumption that homeowners will adopt PV systems only when their cost falls below that of utility supplied power. Only with such cheap PV systems would conventional, inefficient use of electricity with correspondingly large PV systems make much sense.

9. "Alternative Energy Sourcebook" (Ukiah, Calif.: Real Goods Trading Corporation, 1990).

10. D. Carlson, "Photovoltaic Technologies for Commercial Power Generation," *Annual Review of Energy* 15 (1990): 85–98.

11. S. Verchinski, interview at Solar Electric Systems, Albuquerque, N.M. (August 15, 1989).

12. B. Summers, interview at Trace Engineering, Arlington, Wash. (July 28, 1989).

13. W. Dankoff, interview at Flowlight Solar Power, Santa Cruz, N.M. (August 16, 1989).

14. L. Garrett, Sun Amp, Scottsdale, Ariz. (August 9, 1989).

15. S. Willey, interview at Backwoods Solar Electric, Sandpoint, Idaho (July 26, 1989).

16. Ibid.

17. An example of each from site visits and personal interviews in July and August 1989, can be found in Earth Lab, Willits, Calif.; Mendocino Power Company, Ukiah, Calif.; and Sunelco, Hamilton, Mont.

18. Even though there are no robust statistics to support this statement, it seems to be borne out by the location of businesses, by the advertising, reader letters, and other contents of *Home Power Magazine* (Ashland, Ore.), and by activity at the Renewable Energy Fair held annually in Amherst, Wis. Other sources include a 1991 Real Goods survey, Real Goods Trading Corporation, *The Real Stuff* 1, no. 2 (July 1991): 3, indicating only 6 percent of home power customers were from the eastern United States. This result is likely to be biased given the location of Real Goods itself, as the home power business was overwhelmingly mail order at the time and Real Goods was the largest such vendor. In addition, California was, at the time, one of the only states in which sales were vigorous enough to support a good number of retail outlets in addition to the mail order businesses.

19. See, for example, activity in Connecticut noted in Andi Rierden, "Homeowners Revive Interest In Solar Power," *New York Times*, (August 25, 1991), section 12, p. 1. See also the national tour of home power homes organized for October 16, 1993, by Real Goods, *Real Goods News* (October 1993); this tour included one or more homes in each of forty states. And, see Potts, *The Independent Home.*

20. This generalization, again, seems to be amply borne out by ethnographic data (Tatum, "The Home Power Movement") and by the contents of *Home Power Magazine.* To be precise and quantitative, more detailed analysis would be required. At this writing, the closest thing to a detailed survey offering this kind of characterization is found in a survey done by the Real Goods Trading Corporation (July 1991).

21. The communities referred to are near Sandpoint, Idaho; Tonasket, Washington; Pinehurst, Oregon; Garberville, California; and Davenport, California.

22. Tatum, "The Home Power Movement," p. 99–107.

23. Laura Nader and S. Beckerman, "Energy as It Relates to the Quality and Style of Life," *Annual Review of Energy* 3 (1978): 1–28.

24. J. Holdren, G. Morris and I. Mintzer, "Environmental Aspects of Renewable Energy Sources," *Annual Review of Energy* 5 (1980): 241–291. Organization for Economic Cooperation and Development, "Environmental Impacts of Renewable Energy" (Paris: Organization for Economic Cooperation and Development, 1988).

25. J. Ogden and R. Williams, "Solar Hydrogen: Moving Beyond Fossil Fuels," (Washington, D.C.: World Resources Institute, 1989).

26. For a more typical image of consumer behavior, see Willit Kempton and L. Montgomery, "Folk Quantification of Energy," *Energy—The International Journal* 7 (1982): 817.

27. See, for example, the compilation of renewable energy projects in Kathleen Courrier et al., eds., *Shining Examples* (Washington, D.C.: Center for Renewable Resources, 1980); or Nancy Rader, "The Power of the States: A Fifty-State Survey of Renewable Energy" *Public Citizen*, (June 1990).

28. Langdon Winner makes a similar point in arguing that we should not accept economic standards uncritically when we talk about energy. Langdon Winner, *The Whale and the Reactor* (Chicago: University of Chicago Press, 1984), p. 54.

29. Greenwald, "Here Comes the Sun," pp. 84–85.

30. See, for example, Everett M. Rogers, *Diffusion of Innovations* (New York: The Free Press, 1962).

31. See, for example, Neil J. Smelser, *Theory of Collective Behavior* (New York: The Free Press, 1983).

32. The discussion that follows relies heavily on a previously published article; Tatum, "Technology and Values."

33. Albert Borgmann, *Technology and the Character of Contemporary Life* (Chicago: University of Chicago Press, 1984), p. 41–42.

34. Weinberg's discussion of the "technological fix" comes close to an explicit statement of a pattern for technology development consistent with the device paradigm. See Alvin Weinberg, "Can Technology Replace Social Engineering?" reprinted in *Technology and the Future*, Albert Teich, ed. 5th edition (New York: St. Martin's Press, 1990), pp. 29–38.

35. Borgmann, *Technology*, p. 42.

36. Ruth Schwartz Cowan, *More Work for Mother* (New York: Basic Books, 1983).

37. Borgmann, *Technology*, p. 51.

38. Again, disburdenment is not necessarily to be denigrated. As a reviewer has pointed out, that "frozen dinner might not only seem like a liberation to the wife, [it] might be precisely the thing which will allow her to express some value more central to her than pleasing her husband's sense of a 'full bodied Christmas' by making cooking her focal activity—such as reading a book or teaching a class." Granted, where the book is an orienting engagement, it may be appropriate to buy the frozen dinner. Where the traditional dinner is the thing, however, the frozen "equivalent" clearly does not do the job and perhaps should not be embraced; its disburdening effect should not blind us to its other implications. Neither Borgmann nor I intend to offer an invariant classification of particular technologies as either things or devices, nor does either of us suggest that the disburdening benefit of technology should always be eschewed.

39. Borgmann, *Technology*, p. 211.

40. Ibid, p. 223.

41. Ibid, p. 240.

42. Ibid, p. 231.

43. Numerous analyses of the apparent demise of nuclear power in this country may be instructive here. See, for example, Joseph G. Morone and Edward J. Woodhouse, *The Demise of Nuclear Energy: Lessons for Democratic Control of Technology* (New Haven, Conn.: Yale University Press, 1989).

44. Lawrence Tribe, "Technology Assessment and the Fourth Discontinuity: The Limits of Instrumental Rationality," *Southern California Law Review* 46 (1973): 617–660.

Chapter 7. Where Does This Leave Us?

1. Michael Agar uses the apt term "professional stranger" to refer to the ethnographers that I have particularly in mind. See Michael Agar, *The Professional Stranger* (New York: Academic Press, 1980).

2. Useful insights may be offered here by focusing on the symbolic value of particular technologies at particular times. See, for example, James J. Flink, *The Car Culture* (Cambridge, Mass.: MIT Press, 1975).

3. I am using the language of the emerging field of study of science, technology and society. See, for example, Bruno Latour, *Science in Action* (Cambridge, Mass.: Harvard University Press, 1987) or Wiebe E. Bijker, Thomas P. Hughes, and Trevor Pinch, eds., *The Social Construction of Technological Systems* (Cambridge, Mass.: MIT Press, 1987).

4. Even without ethnographic work specifically targeting energy-related possibilities, any number of classic works describing a wide range of values, traditions, and subcultures within the United States could be used as a starting point in generating a wide range of different energy scenarios. Just to scratch the surface of what is available, see A. J. Vidich and J. Bensman, *Small Town in Mass Society* (Princeton, N.J.: Princeton University Press, 1968); Carol Stack, *All Our Kin: Strategies for Survival in a Black Community* (New York: Harper and Row, Publishers, 1974); Robert N. Bellah et al., *Habits of the Heart* (Berkeley: University of California Press, 1985); or Duane Elgin, *Voluntary Simplicity: Toward a Way of Life That Is Outwardly Simple, Inwardly Rich* (New York: William Morrow and Company, 1981). With specific reference to energy, see also Energy Research and Development Administration, *Solar Energy in America's Future*, prepared by Stanford Research Institute International for the Energy Research and Development Administration, March 1977, DSE-115/1, Chapter 5, "Broader Issues," by Willis Harman et al.; Paul C. Stern and Elliot Aronson, *Energy Use: The Human Dimension*, prepared by the Committee on Behavioral and Social Aspects of Energy Consumption and Production, National Research Council (New York: W. H. Freeman and Company, 1984).

5. Edward S. Cassedy and Peter Z. Grossman, *Introduction to Energy: Resources, Technology, and Society* (Cambridge: Cambridge University Press, 1990), p. 19.

6. For a survey, see Daniel J. Fiorino, "Citizen Participation and Environmental Risk: A Survey of Institutional Mechanisms," *Science, Technology, and Human Values* 15, no. 2 (Spring 1990): 226–243; and Frank N. Laird, "Participatory Analysis, Democracy, and Technological Decision Making," *Science, Technology, and Human Values* 18, no. 3 (Summer 1993): 341-361.

7. For an introduction to constructive technology assessment, see Johan W. Schot, "Constructive Technology Assessment and Technology Dynamics: The Case of Clean Technologies," *Science, Technology, and Human Values* 17, no. 1 (Winter 1992): 36–56. For a review of international instututional arrangements for doing constructive technology assessment, see Norman J. Vig (Department of Political Science, Carleton College), "Parliamentary Technology Assessment in Europe: Comparative Evolution," paper delivered at the September 3–6, 1992, annual meeting of the American Political Science Association. For consideration of the need for constructive technology assessment and for thoughts on where it may be most needed and how it can be pursued in theory and practice, see Jesse S. Tatum, "Reflections on the Theory and Practice of Constructive Technology Assessment," *Proceedings of the Eighth Technology Literacy Conference of the National Association for Science, Technology, and Society*, ed. Dennis Cheek (Bloomington, Ind.: Educational Resources Information Clearing House for Social Sciences and the Social Studies, 1993), pp. 22–32.

8. Schot, ibid.

9. Lewis Mumford, *The Pentagon of Power* (New York: Harcourt Brace Jovanovich, 1967), pp. 172–173.

10. Herbert Marcuse, *One Dimensional Man* (Boston: Beacon Press, 1964).

11. Robert Green Ingersoll, "Some Reasons Why" (1896), quotation in John Bartlett, *Familiar Quotations*, 15th Ed., ed. Emily M. Beck (Boston: Little, Brown and Company, 1980), p. 615.

12. Prospero, in Shakespeare's *The Tempest*, commands awesome and fantastic powers for a time:

... I have bedimm'd
The noontide sun, call'd forth the mutinous winds,
And 'twixt the green sea and the azur'd vault
Set roaring war; to the dread rattling thunder
Have I given fire and rifted Jove's stout oak
With his own bolt; the strong-bas'd promontory
Have I made shake and by the spurs pluck'd up
The pine and cedar; graves at my command
Have wak'd their sleepers, op'd, and let 'em forth
By my so potent Art. . . .

Yet when the powers he has exercised have achieved their just and proper cause, their exercise is set aside. In his wisdom, Prospero somehow recognizes even at the apogee of success that the powers of nature he has indeed exercised have yet never been properly his:

... But this rough magic
I here abjure; and when I have requir'd
Some heavenly music (which even now I do)
To work mine end upon their senses that
This airy charm is for, I'll break my staff,
Bury it certain fathoms in the earth,
And deeper than did ever plummet sound
I'll drown my book.

From William Shakespeare, *The Tempest*, The Blackfriars Shakespeare, ed. Leonard Nathanson (Dubuque, Iowa: Wm. C. Brown Company Publishers, 1969), pp. 48–49.

13. In this sense, they would be consistent with recommendations made by Borgmann and others, and might be considered a form of the "deictic" discourse that Borgmann urges—see Albert Borgmann, *Technology and the Character of Contemporary Life* (Chicago: University of Chicago Press, 1984), p. 178, etc.

14. Roger Hausheer, "Introduction," in Isaiah Berlin, *Against the Current: Essays in the History of Ideas* (New York: Penguin Books, 1982), p. liii.

15. Amory Lovins, *Soft Energy Paths: Toward a Durable Peace* (Cambridge, Mass.: Ballinger Publishing Company, 1977), e.g., p. 59.

16. Paul Feyerabend, *Science in a Free Society* (London: Verso Editions/NLB, 1978), p. 80.

17. See, for example, David Orr, "U.S. Energy Policy and the Political Economy of Participation," *Journal of Politics* 41, no. 4 (November 1979): 1028–1056.

18. Before Amendment 17 (1913) required direct election of U.S. senators, the U.S. Constitution provided that "The Senate of the United States shall be composed of two Senators from each state, *chosen by the Legislature thereof* . . . [emphasis added]." The president is, of course, chosen by the Electoral College (not by direct popular ballot) even today.

Index